Science and Society

Eric S. Swanson

Science and Society

Understanding Scientific Methodology,
Energy, Climate, and Sustainability

 Springer

Eric S. Swanson
Department of Physics and Astronomy
University of Pittsburgh
Pittsburgh, PA, USA

ISBN 978-3-319-21986-8   ISBN 978-3-319-21987-5   (eBook)
DOI 10.1007/978-3-319-21987-5

Library of Congress Control Number: 2015947274

Springer Cham Heidelberg New York Dordrecht London
© Springer International Publishing Switzerland 2016
This work is subject to copyright. All rights are reserved by the Publisher, whether the whole or part of the material is concerned, specifically the rights of translation, reprinting, reuse of illustrations, recitation, broadcasting, reproduction on microfilms or in any other physical way, and transmission or information storage and retrieval, electronic adaptation, computer software, or by similar or dissimilar methodology now known or hereafter developed.
The use of general descriptive names, registered names, trademarks, service marks, etc. in this publication does not imply, even in the absence of a specific statement, that such names are exempt from the relevant protective laws and regulations and therefore free for general use.
The publisher, the authors and the editors are safe to assume that the advice and information in this book are believed to be true and accurate at the date of publication. Neither the publisher nor the authors or the editors give a warranty, express or implied, with respect to the material contained herein or for any errors or omissions that may have been made.

Cover illustration: Detroit Industry, 1933 (detail of the north atrium wall, fresco) by Diego Rivera/ Detroit Institute of Arts/Bridgeman Images

Printed on acid-free paper

Springer International Publishing AG Switzerland is part of Springer Science+Business Media (www. springer.com)

To the world's seven billion scientists

Praise for Science and Society

"Eric Swanson has managed to blend rigorous scientific and physical concepts with assessments of real problems faced by society - many of which have important scientific components. Building scientific intuition and wider literacy about current challenges will be essential to making better decisions in the future, and this book is an excellent step forward."

Gavin Schmidt, director of the Goddard Institute for Space Studies, author of "Climate Change: Picturing the Science", and co-founder of the award winning blog, "RealClimate.org"

"Anyone who thinks society can be managed without science should think again, or better: read this book. Eric Swanson explains how science permeates society, and with simple examples of the scientific process he shows its special power in dealing with the natural world. This is a must read for the world's seven billion scientists."

F.E. Close, OBE, Oxford University, author of "Half-Life: The Divided Life of Bruno Pontecorvo, Physicist or Spy", "The Infinity Puzzle", "Neutrino", and several others

"After working for more than five years consulting on various energy-related projects for two Department of Defense facilities, I wish I had this book available to recommend to energy managers and other decision makers. This book will be valuable beyond the classroom."

Alex R. Dzierba, Chancellor's Professor of Physics Emeritus, Indiana University

Preface

We live in the age of science. Almost every facet of our society is influenced by scientific discoveries, many of which have happened in the past two centuries. It is foolish to imagine that we can manage this society, and all of the changes it brings about, without some knowledge of science and the way it works. This book is dedicated to bringing this knowledge.

The book is primarily directed toward nonscience undergraduate students who seek to expand their knowledge of the rules of science that underpin our existence, our impact on nature, and nature's impact on us. Of course anyone else who is curious about these matters can benefit from it as well! I assume that the typical reader will have taken high school physics and chemistry – but forgotten most of it. Thus, I remind the reader of basic things as we go. The use of mathematics will be kept to a minimum, but part of the purpose of this book is to show that simple applications of arithmetic (usually multiplication and division, but sometimes an exponential or a logarithm) permit much knowledge to be gained. Thus, we will not shy away from arithmetizing our considerations – after all, part of the scientific method is being quantitative.

The prime purpose of the book is to explain how the scientific process works and the power it brings to dealing with the natural world. The main applications will be to understand scientific results that flow from the media and to develop a rational, fact-based assessment of energy and resource policy. In this regard, I admit to being a data geek. This is good, because part of adopting a scientific worldview is backing up what you are saying with numbers. I'm going to be saying a lot, so this book is chock full of tables and figures. If you find it overwhelming after a while, skip the numbers, but try to take the message in.

The book is designed to accompany a course that is taught in approximately 30 lecture hours. Thus, its length has purposefully been kept in check, although it is still a bit too long to cover in 30 hours. Of course a student can easily read it during the course of a term!

Occasional sentences will appear in boxes throughout the text. These are take-away points – if you remember anything at all of this book, it should be these messages.

You will find a section called "Preliminaries" right after the List of Tables. You should read this carefully if it has been more than a year since you have done any physics or chemistry. You will find reminders on scientific notation and SI prefixes, the units used in this book, and some common physical constants that appear throughout the text. There is also a discussion of something called "dimensional analysis." This is a fancy way of saying that units must work out properly. Don't ignore this section – it can be surprisingly useful and powerful! Finally, you will find short tutorials on using exponentials, logarithms, and some basic chemistry.

Chapters end with a list of important terminology and important concepts. These are *not* replacements for reading the chapter; rather, they are meant to guide you in case you missed something. There are also exercises at the end of each chapter. These are typically conceptual, but sometimes some simple arithmetic is required. If you are stumped, check "Preliminaries" for general help and "Problem Solutions" at the end of the book for a sampling of worked problems. Finally, chapters typically feature many examples; these either elaborate on a point or ask something numerical.

Lastly, I would like to thank Charlie Jones and Liam Swanson for insightful comments on Chaps. 1 and 10, Max Swanson for superb proofreading, Sam Schweber for pointing me to the intriguing story of Fritz Houtermans, Josef Werne and Gavin Schmidt for discussions on subtleties in climate modeling, Gregory Reed for discussions on power systems, and Jim Richardson for permission to reproduce his stunning photograph of Easter Island.

Pittsburgh, USA   Eric S. Swanson

Contents

List of Figures

List of Tables

Preliminaries

Information that is used throughout this book is collected here for easy reference. The most important things you should learn before starting the book are *scientific notation* and *SI prefixes*. The physical constants are numbers that appear in various places in the book. They are placed here for your convenience and you need not worry about their definitions right now. Lastly, *units* are required to make sense of measurements and *dimensional analysis* is a powerful tool that uses units. Review the units shown below – you are likely familiar with most of them – and don't worry about dimensional analysis if you find it confusing.

Scientific Notation

René Descartes invented a convenient way of writing large or small numbers that is used throughout the scientific world. The idea is to represent repeated multiplication with a *power*. Thus, $3^4 = 3 \cdot 3 \cdot 3 \cdot 3$. Scientific notation focuses on powers of 10; hence,

$$3.14 \cdot 10^4 = 3.14 \cdot 10 \cdot 10 \cdot 10 \cdot 10 = 31400 \tag{1}$$

and

$$3.14 \cdot 10^{-2} = 3.14 \cdot \frac{1}{10} \cdot \frac{1}{10} = 0.0314. \tag{2}$$

Système International Prefixes

A system of prefixes is used in the Système International (SI) to indicate powers of 10. The prefixes that appear in the book are listed here.

Prefix	Symbol	Meaning
femto	f	10^{-15}
pico	p	10^{-12}
nano	n	10^{-9}
micro	μ	10^{-6}
milli	m	10^{-3}
kilo	k	10^{3}
mega	M	10^{6}
giga	G	10^{9}
tera	T	10^{12}
peta	P	10^{15}
exa	E	10^{18}

Constants

Constants of nature that appear in the text are gathered here. These constants allow the conversion of one quantity into another while preserving units.

Name	Symbol	Value
Speed of light	c	$2.998 \cdot 10^{8}$ m/s
Planck's constant	h	$6.626 \cdot 10^{-34}$ J s
Boltzmann's constant	k_B	$1.381 \cdot 10^{-23}$ J/K
Avogadro number	N_A	$6.022 \cdot 10^{23}$/mol
Newton's constant	G	$6.674 \cdot 10^{-11}$ m^3 kg^{-1} s^{-2}
Stefan-Boltzmann constant	σ	$5.670 \cdot 10^{-8}$ W/(m^2 K^4)
Permittivity of the vacuum	ϵ_0	$8.854 \cdot 10^{-12}$ A^2 s^4 kg^{-1} m^{-3}
Permeability of the vacuum	μ_0	$4\pi \cdot 10^{-7}$ kg m s^{-2} A^{-2}

Système International Units

The Système International (also called the *metric system*) is a standardized collection of units used to measure things. SI will be used throughout this book. A few other common units will also appear. These are listed below.

Name	Symbol	Quantity	Conversion
gram	g	mass	–
kilogram	kg	mass	$1000\,g$
tonne	t	mass	$1000\,kg$
meter	m	length	–
kilometer	km	length	$1000\,m$
light year	ly	length	$9.41 \cdot 10^{18}\,m$
angstrom	Å	length	$10^{-10}\,m$
hectare	ha	area	$10{,}000\,m^2$
liter	L	volume	$1/1000\,m^3$
us gallon	gal	volume	$3.785\,L$
barrel	b	volume	$159\,L$
second	s	time	–
hertz	Hz	cycles per second	$1/s$
joule	J	energy	$1\,kg\,m^2/s^2$
electron volt	eV	energy	$1.6021 \cdot 10^{-19}\,J$
calorie	Cal	energy	$4184\,J$
british thermal unit	btu	energy	$1055.87\,J$
watt	W	power	$1\,kg\,m^2/s^3$
horsepower	hp	power	$746\,W$
sievert	Sv	radiation dose	–
degree Celsius	C	temperature	–
degree Kelvin	K	temperature	$C-273.15$
degree Fahrenheit	F	temperature	$32 + 1.8\,C$
ampere	A	current	$6.241 \cdot 10^{18}$ electrons/s

Units and Dimensional Analysis

Science has a way to formalize the expression "don't compare apples and oranges." Specifically, measurable things must be specified in some units, which means they must be compared to a standard. Thus, we do not say we weigh "150" or that it is "300" to Washington, but rather that we weigh 150 pounds and it is 300 km to Washington.

When quantities with units are combined together (in a formula), the resulting expression must carry the correct units. This can be a powerful tool.

> Ex. A simple pendulum is a bob of mass m hanging from a swing of length ℓ. It oscillates with a fixed period, called T. Gravity is the only force acting on the bob, and it is characterized by the acceleration due to gravity, called g. Because the only quantities that describe the pendulum are m, ℓ, and g, it must be true that the period can only depend on the numerical values of m, ℓ, and g. We say that the period is a *function* of these three quantities and write this as:
>
> $$T = T(m, \ell, g). \qquad (3)$$
>
> Because the units of these quantities are [kg], [m], and [m/s^2], the only way to obtain a time is if the dependence goes like
>
> $$T = \text{constant}\sqrt{\ell/g}. \qquad (4)$$
>
> This is in fact the correct formula (the constant is 2π).[1]

> Ex. Consider the area of a right triangle with side lengths given by a, b, and the hypotenuse c. Since the units of these lengths are [m] and the units of area are [m^2], the area of the triangle must depend on `side lengths` times `side lengths`. This is written as
>
> $$A = A(a^2, b^2, c^2, ab, ac, bc). \qquad (5)$$
>
> Let us call one of the acute angles β. Then, the trigonometric relationships $a = c\sin\beta$ and $b = c\cos\beta$ can be used to rewrite A as
>
> $$A = A(c^2\sin^2\beta, c^2\cos^2\beta, c^2, c^2\sin\beta\cos\beta, c^2\sin\beta, c^2\cos\beta). \qquad (6)$$
>
> Since the area has units of [m^2] and angles don't have units, A must be proportional to c^2:
>
> $$A = c^2 f(\beta) \qquad (7)$$

---

[1] This simple model of a pendulum is for small oscillations and ignores air resistance, twist in the swing, etc., which all introduce more variables.

where f is some function[2] of the angle β. Notice that the definitions of sides a and b can be flipped so it does not matter which angle is used.

Now divide the triangle into two smaller ones as indicated in the figure. The smaller triangles have hypotenuses of length a and b and the areas must sum to give $c^2 f(\beta)$:

$$c^2 f(\beta) = a^2 f(\beta) + b^2 f(\beta). \tag{8}$$

Dividing by f gives
$$a^2 + b^2 = c^2, \tag{9}$$

which is the famous Pythagorean theorem – all from simple dimensional analysis!

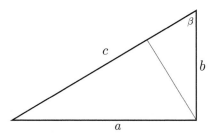

Dimensional proof of the Pythagorean theorem

Exponentials and Logarithms

The exponential and logarithmic functions appear in Chap. 9 because they describe things like population growth. Two notations are in common use for the exponential:

$$y = \exp(x) \tag{10}$$

or

$$y = e^x. \tag{11}$$

The symbol e refers to *Euler's number*, $2.7182818\ldots.$ This number appears in many places, but perhaps its simplest application is in solving the problem

$$\text{slope of function} = \text{function.} \tag{12}$$

---

[2] Noting that a right triangle can be completely specified by giving one angle and the length of the hypotenuse allows one to write this equation immediately.

It is not difficult to derive a formula for e. If we call the function above f, then we seek the value of f at unity, given that $f(0) = 1$. Thus, $e = f(1)$. The slope of f at a point slightly above zero is given by $(f(\delta) - f(0))/\delta$. Remember that this must be $f(0)$ and $f(0) = 1$. We consider δ to be very small so that the slope formula is accurate. The conclusion is $f(\delta) = (1 + \delta)$. The same argument can be used to get $f(2\delta) = f(\delta) \cdot (1 + \delta) = (1 + \delta)^2$. Now repeat to obtain

$$f(n\delta) = (1 + \delta)^n \tag{13}$$

and set $\delta = 1/n$. The result is

$$e = f(1) = \left(1 + \frac{1}{n}\right)^n. \tag{14}$$

This formula becomes more accurate as n gets larger. Thus, for $n = 10$, it gives $e \approx 2.59379$; for $n = 1000$, $e \approx 2.7169$; and for $n = 1000000$, $e \approx 2.71828$.

Using the exponential function amounts to multiplying e by itself many times. For example,

$$\exp(3) = e^3 = e \cdot e \cdot e \approx 20.0855. \tag{15}$$

In practical situations, you should just use a calculator or a calculator app to get this number.

The *logarithm* is the inverse of the exponential and is denoted "ln," or sometimes "$\log_e$" or "log." Thus, if $\exp(3) = 20.0855$, then $\ln(20.0855) = 3$ (approximately). Again, you should use a calculator to compute logarithms when they are needed. Detailed examples that use these functions are given at the end of the text.

Mass and Volume of Simple Substances

In Chap. 4, you will learn that the mass of a proton is about $1.67 \cdot 10^{-27}$ kg. This means that about $6 \cdot 10^{23}$ protons will have a mass of 1 g. The last number is called the *Avogadro number* (denoted N_A in the table of constants above). The concept of a *mole* is often used in this context. A mole (abbreviated "mol") is defined as the amount of any substance that contains N_A objects (i.e., atoms, molecules, etc.). Although moles of substances will be referred to in the text, you really don't need to use this idea at all – just remember that N_A protons weigh 1 g and N_A is a fixed large number.

Say you want to know how much a mole of carbon dioxide weighs. The chemical formula for carbon dioxide is CO_2. To get the answer, use a periodic table or app to determine that carbon consists of six protons and six neutrons.

Since a neutron has about the same mass as a proton, N_A carbon atoms weigh 12 g. Oxygen consists of eight protons and eight neutrons, so N_A pairs of oxygen atoms (recall we have O_2) weigh 32 g. The total is then $12 + 32 = 44$ g.

There are several subtleties in this example: (i) neutrons have slightly higher masses than protons; (ii) electrons have mass; (iii) when neutrons and protons stick together to make an atomic nucleus, the total mass is slightly reduced; (iv) when atoms stick together to make molecules, the total mass is slightly reduced; (v) when molecules stick together to make a gas, liquid, or solid, the total mass is slightly reduced; (vi) different numbers of neutrons can be present in a given element, which throws the calculation off. Except for point (vi), all of these effects are small, and we ignore them in this book.

Knowing the number of molecules in a gas of a simple substance allows one to determine its volume. This is achieved with the *ideal gas law* of high school chemistry fame:

$$PV = nRT. \tag{16}$$

Here n is the number of moles of gas and R is the *gas constant* given by $R = 0.08206 \, \text{L} \cdot \text{atm}/(\text{mol} \cdot \text{K})$. To use this equation, you need to measure volume in liters, temperature in degrees Kelvin, and pressure in atmospheres. The formula tells us that the volume of one mole of any gaseous substance at $0\,\text{C}$ (this is 273.15 K) is

$$V = \frac{nRT}{P} = 1 \text{ mol} \cdot 0.08206 \, \frac{\text{L atm}}{\text{mol K}} \cdot 273.15 \text{ K} \cdot \frac{1}{1 \text{ atm}} = 22.4 \text{ L}. \tag{17}$$

At higher temperatures, the volume would be larger, as specified by the ideal gas law. As with molar masses, this result is an approximation that ignores deviations from the ideal gas law that occur for all real substances. Again, we will not be concerned with these small corrections.

1

What Is Science?

"[Natural] philosophy is written in this grand book – I mean the universe – which stands continually open to our gaze, but it cannot be understood unless one first learns to comprehend the language and interpret the characters in which it is written. It is written in the language of mathematics, and its characters are triangles, circles, and other geometrical figures, without which it is humanly impossible to understand a single word of it; without these, one is wandering around in a dark labyrinth."

— Galileo Galilei, *The Assayer*

Defining science is not as simple as one might think. In retrospect this is not surprising; it is difficult to *define* many things. We all know happiness or love or evil when we see it, but defining these concepts is difficult indeed. Even a physical entity like a duck is hard to define. The dictionary definition, "A waterbird with a broad blunt bill, short legs, webbed feet, and a waddling gait" seems unsatisfactory. This description of a few duck-like attributes does not seem to capture the essence of a duck.

But should one expect more of a definition? If a thing is the sum of its attributes then the best one can do is list these attributes. This is certainly a practical point of view, and above all, science is practical.

1.1 Modes of Logic

At an intuitive level, we feel that science involves a rational examination of the natural world. But most scientific effort seeks to move beyond data collection by building a unifying description of the essential phenomena. Such a description is called a *model* or a *theory*. Since model building is an example of inductive logic it is useful to briefly review the differences between deduction and induction.

Deductive reasoning or deductive inference is the derivation of a statement that must be true given a set of premises (assumptions). For example:

© Springer International Publishing Switzerland 2016
E.S. Swanson, *Science and Society*,
DOI 10.1007/978-3-319-21987-5_1

```
All Canadians like hockey
Jane is a Canadian

Therefore Jane likes hockey
```

Notice that the veracity of the premises is not at issue here; only whether the conclusion logically follows if the premises are true.

Deduction is not particularly useful; deducing an absolutely true statement requires strong premises. It seems that the more interesting part of the deductive process is in forming the premises. How does one know, for example, that Jane is a Canadian? Maybe she claims she is, or you know she was born there, or she has a Canadian passport, but none of these *prove* she is Canadian. Making this statement then relies on extrapolating from evidence to hypothesis.

Deduction is dull. Induction is interesting.

This extrapolation is a form of *inductive* reasoning. Induction relies on extrapolating a series of observations. For example, the sun has risen every day of your life, and you fully expect it to rise tomorrow. There is no *logical* justification for this expectation. However, its feasibility certainly seems justified. Notice that the degree of certainty of an inductive conclusion can change with time. For example, if the sun continues to rise for 100 years, one would presumably have more confidence in the hypothesis. At a deeper level, if one knows about planetary dynamics and the rotation of the Earth, then one can state with certainty that the sun will rise barring apocalyptic events.

It is clear that all of the important advances in science have been inductive: Newton's universal gravitation, the atomic paradigm, and Einstein's (1879–1955) theory of relativity have all extrapolated from a small set of observations to general properties of the universe.

Although inductive reasoning seems entrenched in science, and in fact, in all human activity, Scottish philosopher David Hume (Fig. 1.1) argued that inductive reasoning cannot be justified and is more of an animal instinct. Beginning with his *A Treatise of Human Nature* (1739), Hume strove to create a naturalistic "science of man" that examined the psychological basis of human nature. In stark opposition to the rationalists who preceded him, most notably Descartes, he concluded that desire, rather than reason, governed human behavior: "Reason is, and ought only to be the slave of the passions".

Hume notes that induction relies on an assumed *uniformity of nature*, i.e., what you experience here and now applies to there and then. Of course, such an assumption is not logically provable, as one cannot prove a negative. However most rational people do not go about life suspecting that apples will suddenly levitate or air will become poisonous.

Figure 1.1: David Hume (1711–1776) was a Scottish philosopher, historian, economist, and essayist known especially for his philosophical empiricism and skepticism.

While Hume's problem of induction has caused much consternation in philosophic circles, it has not had much effect on science. In fact Hume's concept of uniformity of nature has been promoted to a pair of fundamental principles of modern physics. These are

(i) the results of all experiments do not depend on *when* they are conducted.

(ii) the results of all experiments do not depend on *where* they are conducted[1]

In terms of eliminating assumptions, it appears that nothing has been gained. But the mathematization of physics permits a surprising and powerful extension of these principles. Namely, both principles are an example of ***symmetry*** in nature. Symmetry is not important in the remainder of this book, so I'll just note that symmetry means that an object looks the same after you do something to it (like rotate it or reflect it).

In mathematical physics symmetries of nature imply ***conservation laws***. A conservation law is a statement that some quantity does not change in time. Perhaps this is a surprising concept – after all everything in our common experience changes with time: people age, buildings decay, mountains wash into the sea, the sun slowly burns out.

Modern physics postulates a small set of quantities that do not change in time. These are things like the charge of the electron, the mass of the top quark, the total electric charge, and a collection of other more or less esoteric quantities.

---

[1]Of course one assumes that everything else is kept equal.

There are also quantities associated with motion that are conserved: energy, momentum, and angular momentum. Each of these is associated with a symmetry that underpins it.

Our uniformity principles are really symmetries of nature and the connection between symmetry and conservation laws yields the equivalent statements to rules (i) and (ii) above:

(i') energy is conserved

(ii') momentum is conserved.

Although these laws can never be proven since that requires examining all experimental outcomes in all of space and in all of time, *no exception has ever been observed*. The accumulation of thousands of years of experience and millions of experimental results have led to a fair degree of confidence (to put it mildly) in the (logically unjustified) assumption of the uniformity of nature.

We will discuss laws in more detail shortly, but first must address an important issue that exists with induction.

1.2   Modelling and Occam's Razor

So far the induction we have discussed is of the form, "All bees I have ever seen are yellow, so bees are yellow". While this is fine, it is not a particularly powerful method for building models of nature, which typically involves a lot more guess work.

Philosophers of science often refer to another sort of induction called *inference to the best explanation* (or IBE for short). *This* is what we mean by model building. An example of IBE is

```
A dead mouse is found on the door step.
The cat is looking pleased with itself.

The cat killed the mouse.
```

Clearly the conclusion is not a logical consequence of the premises, however reasonable it may be. Perhaps less obvious is that a principle of parsimony has been invoked, namely the inference fits the observations and is not overly elaborate (this is why the word 'best' appears in IBE). For example, one could postulate that the cat looks pleased because it just had a long nap and that the dead mouse was deposited by a visiting alien who was momentarily interested in mouse anatomy. This would explain the data equally well, but is certainly not as parsimonious.

The work of scientists is largely the form of such educated guesses. A famous example is provided by Darwin's theory of evolution. Darwin supported his model by noting that various observations concerning the animal world are difficult to understand if species were created separately but are simple to understand if they have evolved from common ancestors. Why, for example, would mammals all share the same basic furry, four-legged body form if God had created them independently? It seems unimaginative. Alternatively, a basic body form is natural if mammals have descended from a common ancestor.

The concept of building a 'best' model is often phrased in terms of *Occam's razor*. Occam's razor, originally stated by William of Ockham (Fig. 1.2), stipulates that the simplest possible explanation for a phenomenon is to be preferred. This seems reasonable enough, and has been universally agreed upon since the ancient Greeks.[2]

Figure 1.2: William of Ockham (c. 1287–1347). English Franciscan friar, scholastic philosopher, and theologian.

Credit: User:Moscartop/ Wikimedia Commons/ CC-BY-SA-3.0.

Nevertheless the application of Occam's razor is not without problems. Who, for example, is to decide what is 'simpler'? And what does nature care of our paltry ideas of simplicity? It seems clear that the choice of a preferred model depends on many extraneous things such as the experience, cultural preferences, and prejudices of the modeler.

Another problem with the application of Occam's razor is that the modeler must choose which set of observations with which to compare her model. It is all too human to simply ignore inconvenient data! For example, early models of the Earth's structure that postulated tectonic motion were rejected because the idea of continents that moved seemed implausible. But this position was reversed as

---

[2]Consider the following. Aristotle: "We may assume the superiority, all things being equal, of the demonstration which derives from fewer postulates or hypotheses." Ptolemy: "We consider it a good principle to explain the phenomena by the simplest hypothesis possible." Duns Scotus: "Plurality is not to be posited without necessity". Isaac Newton: "We are to admit no more causes of natural things than such as are both true and sufficient to explain their appearances". Einstein: "Everything should be kept as simple as possible, but no simpler."

more evidence in favor of the hypothesis was discovered (see Fig. 1.3). Similarly the atomic paradigm was thought to be complex because it implied the existence of particles that had not been observed. Again, the weight of observations, and especially the parsimonious explanation of Brownian motion in terms of atomic interactions by Einstein, turned the tide.[3]

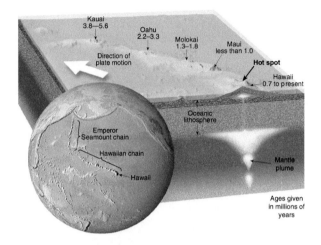

Figure 1.3: Evidence in favor of tectonic motion: J. Tuzo Wilson (1908–1993) realized that the string of Hawaiian islands could be explained by a tectonic plate moving over a hot spot.

Copyright, Tasa Graphic Arts. Reproduced with permission.

A more extreme version of Occam's razor was famously espoused by Paul Dirac.[4] Dirac suggested that nature prefers models that are mathematically beautiful. Given the difficulty in defining beauty, this viewpoint must be considered even more anthropomorphic than Occam's razor.[5] Dirac's beliefs took a somewhat tragic turn late in his life. Dirac was particularly proud of the elegant and compact equation he derived to describe the quantum properties of the electron. However it was soon realized that this equation could not be correct and – in Dirac's view – complicated adjustments had to made to it to bring it into agreement with observations. Because of this he thought his equation wrong and his "life a waste".

Although it appears that the application of Occam's razor must be mired in anthropomorphic muck, a cleaner approach based on *counting parameters* is possible.

---

[3]Brownian motion is the random motion of small particles in a fluid.

[4]Paul Adrien Maurice Dirac (1902–1984) was a British theoretical physicist who made fundamental contributions to the early development of quantum mechanics and quantum field theory.

[5]Anthropomorphism is the attribution of human characteristics to nonhuman entities. In this case the very human notions of simplicity or beauty are being applied to nature.

All models of nature depend on *a priori* unknown quantities called **parameters**. For example, if one is modelling the motion of a satellite through the solar system some of the relevant parameters would be the masses of the sun, Earth, and satellite; the strength of the force of gravity (represented by Newton's constant); and the initial speed of the satellite. Replacing the rather nebulous ideas of simplicity or beauty with 'fewer parameters' removes the anthropomorphic element of Occam's razor from consideration. Indeed, models with fewer parameters are often more complex than those with more parameters, but have the great benefit of being more **predictive**. Predictiveness is a property of **theories**, which are discussed in Sect. 1.3.2.

1.3 Theory, Model, and Law

So far we have been using the terms theory, model, and law in a rather loose way. Let's get more specific.

1.3.1 Scientific Law

A **law** is a compact description of certain empirical properties of nature. It differs from a theory (see below) in that it does not *explain* the observations, it merely summarizes them. Some important laws are

(i) Conservation of Momentum

Every object has a quantity called **momentum** associated with it (momentum is velocity times mass), and the sum of all momenta never changes. You experience this law with every action you take. For example when you throw a ball, the momentum imparted to you is equal in magnitude and opposite in direction to the momentum imparted to the ball. This law is associated with the uniformity of space.

(ii) Conservation of Energy

Every object has a quantity called **energy** associated with it, and the sum of all energy never changes. There are two kinds of energy, one associated with the motion of an object, called **kinetic energy**, and one associated with the forces acting on the object, called **potential energy**. Forms of energy can be converted into each other, but can never be created or destroyed. This law is associated with the uniformity of time.

(iii) The Second Law of Thermodynamics

Every collection of objects has a quantity called **entropy** associated with it. We will discuss entropy more thoroughly in Chap. 4; for now think of it as disorder in the objects that make up a system. Entropy is useful because the second law of thermodynamics states that the entropy of the universe must always increase. Colloquial versions of this law are:

- Perpetual motion machines are impossible.

- You can't win and you can't break even.

- Disorder always increases in the universe.

- Heat cannot flow from a lower temperature to a higher temperature.

Conservation of energy and momentum is thought to be an exact property of nature. The second law of thermodynamics, on the other hand, is true *on average*, i.e., when a large number of particles are present.[6] We will revisit energy and entropy in Chap. 4.

Although a single counterexample to a law will serve to invalidate it, laws encapsulate observations of some subset of nature, and therefore are always useful when referring to that subset. In this sense laws are never wrong, merely more or less useful.

Energy is conserved.

Momentum is conserved.

Entropy increases.

1.3.2 Scientific Theory

Laws are independent of theories. Theories attempt to explain laws and to fit them in a predictive framework. Notice that this is *not* the common colloquial meaning of theory, which is something more like 'hypothesis'. Thus the "Theory of Evolution" is no mere hypothesis; it is a working framework that underpins all of biology and has been verified countless times in experiment and observation.

Examples of scientific theories are (i) the aforementioned theory of evolution, (ii) Newton's theory of gravitation, (iii) Einstein's theory of gravitation, (iv) the atomic paradigm, (v) the cosmological concordance model, (vi) the Standard Model, and (vii) plate tectonics theory.

---

[6]Most laws are known to be approximations to nature. For example Ohm's law posits a simple relationship between voltage, current, and resistance. However this breaks down in all materials at sufficiently high voltage. Diodes, for example, rely on non-ohmic behavior to be useful.

As with laws, theories are invalidated if they disagree with experiment in any instance. For example, it is known that Newton's theory of gravitation is incorrect when motion becomes fast or masses become large. However, the theory is still accurate in all other cases, and it would be foolish to discard it simply because it is not accurate in all cases. Rather, one notes where the inaccuracies occur and moves on.

An important feature of theories is that they are **predictive**, namely they permit the prediction of the results of specific experiments or observations. Since theories generally have far fewer parameters than the observations they describe, predictiveness implies that one can make statements about nature for which one does not know the answer, and hence theories can be tested.

Finally, theories tend to be mathematical (at least in physics). Namely, the ideas of a theory are encapsulated in mathematical equations. The solutions to the equations then represent the predictions of the theory.[7] This attitude was elegantly expressed by the person widely regarded as the first scientist – see the quotation opening this chapter.

We shall return to both of these points shortly.

1.3.3   Scientific Model

A **model** is like a theory in that it is a logical framework intended to represent reality. However theories tend to be mathematical, while models are conceptual. Thus theories are special cases of models and many of the theories mentioned in the previous section are called "models". The word model is also used for *simplified or idealized representations of nature.* For example, a perfect plane may be used as a model of a table top, a sphere may model a ball, or a point mass may model a planet. The point is that real tables, balls, or planets are complex things, and often one is not interested in the fine details that make up a physical object. For example, when computing the orbit of a satellite it is not necessary to know the internal structure of the sun or whether there is an outbreak of sunspots.[8]

Sometimes modelling is forced on us. In the case of the satellite orbit we would be in trouble indeed if the internal structure of the sun were required to make reasonable predictions. Even a simple problem, such as deciding whether to walk or run in the rain requires modelling. In general it is impossible to decide the correct course of action without detailed knowledge of the rainfall to be expected along the route. Thus a model of the rain is required. Let us assume that it falls at a constant rate. It is also convenient to model the person. For now consider the model person to be a flat rectangular cardboard cutout. Imagine the motion of the cutout along its path: it sweeps out a certain volume (see Fig. 1.4) and therefore absorbs the rain contained in that volume. Because the volume swept

---

[7]This was not always the case. Although early attempts at mathematicization exist, it was only with the advent of the Scientific Revolution that this became an accepted standard.

[8]That is to say, it is not necessary when computing a typical orbit. If enormous accuracy is desired then more detailed knowledge of the sun will be required.

by the cardboard is the same regardless of how fast it moves, it does not matter whether one walks or runs.

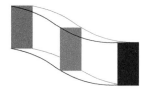

Figure 1.4: A cardboard cutout moving through the rain.

This seems satisfactory, but we must remember that we have made a pretty severe model of a person (namely that he is flat). The model can be tested by *considering limits* (this is a common trick used by scientists to test their models and computations). The model says that it does not matter how fast you go, thus going very slowly (or not at all!) is just as good as running. But only mad dogs and Englishmen stand out in the rain. Something is wrong. A moment's repose reveals the problem: people have thickness and standing out in the rain means that the top of your head and shoulders get wet. To account for this we refine the model by considering a person as a rectangular box of the appropriate size. In this case, the front of the box gets as wet as before, independently of how fast you move, but the top gets less wet as you move faster: it is better to run in the rain.

Models show their true usefulness when they permit predictions. For example, Aristotle (384–322 BC) constructed a model of the universe that postulated four fundamental elements that moved according to their natures. Thus water and earth tended to fall toward the Earth's center, while air and fire rose away from it. However, air remains on Earth and does not escape with ever increasing speed. Aristotle regarded the latter as an absurdity and hence inferred that the universe is finite in size and that it contains an invisible substance that holds the Earth and its atmosphere centered in the cosmos. These dramatic conclusions about the structure of the universe follow directly from his model of the four elements, and were destined to last 2000 years.

A second example from classical Greece is especially compelling. Aristarchus of Samos (310–230 BC) correctly guessed that the Earth, moon, and sun are spheres and that the Earth orbits the sun. It is a simple matter to measure the angular size of the moon and the sun in the sky. It is similarly possible to measure the angle between the moon and the sun when the moon is half full. Finally, assume that a lunar eclipse is caused by the Earth's shadow falling on the moon and that the Earth's shadow is twice as big as the moon itself (this can be estimated by the duration of an eclipse). Then, with the model of Aristarchus and knowledge of the size of the Earth, these few pieces of information are sufficient to yield the sizes of, and the distances to, the sun and moon.

It was a stunning achievement. Suddenly, with the aid of some arithmetic, a few simple observations, and a brave model, it was possible to determine the scale of the solar system and the size of heavenly bodies.

1.4 Defining Science

So far we have seen that science consists of modeling nature with the aid of induction and with the aim of making predictions. Let us then attempt a definition of science:

> Science is the attempt to understand, explain, and predict the world
> we live in. WRONG

This certainly goes in the right direction, but cannot be enough. For example, religion also attempts to make sense of the world, and astrology attempts to make predictions, but neither of these is regarded as science. The distinction can even vary with time. These days the practice of alchemy is thought of as occult or nonsensical, but it was regarded as core science a few hundred years ago.

Perhaps we should step back and list common features of what we regard as science. Science, for example, is often characterized as being experimental. But even this is not adequate; astronomers must rely on observation, not experimentation, when exploring properties of nature.

Let us try again: scientists attempt to extract generalities about nature from the data they collect – namely they build models and theories. But many non-scientific fields model nature and build models. For Karl Popper (Fig. 1.5) the

Figure 1.5: Sir Karl Raimund Popper, (1902–1994) was an Austro-British philosopher and professor at the London School of Economics.

distinguishing feature of scientific endeavor was that it was *falsifiable*, meaning that scientific theories make definite statements about nature that can be tested.

Popper was in part motivated by the burgeoning popularity of Sigmund Freud's new psychoanalytic theory. The problem with Freud's theory was that it could be reconciled with *any* evidence. Popper illustrated this with the following example: a man pushes a child into a river with the intention of murdering him. Another man risks his life to save the child. Freudians can explain both men's behaviour with equal ease: the first man was repressed and the second had achieved sublimation. Popper argued that the use of such concepts made Freud's theory compatible with *any* clinical data and it was thus not falsifiable. Similarly, history and economics have difficulty passing Popper's test since failed predictions can be, and often are, explained away. Popper called theories that provide conceptual frameworks, but which do not provide clear testable predictions, ***pseudo-science***.

Unfortunately, Popper's simple criterion is not as clear as it looks. Experiments often come up with results in disagreement with accepted theory, but scientists do not tend to panic in this situation. Indeed, little progress would be made if scientists discarded their theories at the first sight of trouble. In practice, scientists react in different ways to confounding experiment results:

(i) they reject the experimental result (experimental error is often the source of disagreement)

(ii) they accept the experimental result and modify the theory

(iii) they accept the experimental result and modify assumptions about the system under investigation

(iv) they note the discrepancy, and once the discrepancies build up to an unacceptable level, seek a new theory.

All of these responses occur frequently in science and, depending on circumstances, all are valid. The discovery of Neptune is a famous historical example of this process. Newton's theory explained the orbits of the planets with extraordinary accuracy. However, over time people realized that the orbit of Uranus was slightly, but consistently, in disagreement with Newton's laws. Rather than discard the theory, as Popper would have us do, John Adams (1819–1892) and Urbain Le Verrier (1811–1877) realized that the discrepancy could be explained if another planet existed. Neptune was subsequently discovered at the predicted location.[9]

In a modern example, the radius of the proton has been determined with two different methods that disagree. No one is throwing out the Standard Model or quantum mechanics; rather the discrepancy is generally thought to be an experimental issue or a subtlety in comparing the results.

Is it possible to find a common feature of all things we call 'science' that is not shared by non-science? It appears that the answer is no. In the end it is clear

---

[9]In François Arago's apt phrase, Le Verrier had discovered a planet "with the point of his pen".

that a simple demarcation of science does not exist. It is simply too human an activity. Rather, it seems that science is a loose collection of goals and methodologies. Nevertheless, it does bear certain characteristics that delineate the bulk of its substance: the collection and organization of facts concerning nature, no matter how subject to the frailties and failings of the human mind; the use of induction to construct conceptual frameworks with predictive power; and the use of mathematics to describe the framework and its predictions.

Science is a human activity.

1.5   Progress and Change in Science

Above all, science is utilitarian, and if a theory should produce correct predictions almost all of the time, it is not going to be discarded with the bath water. Scientists will simply note where the theory has failed and move on. Of course too many failures will lead to doubt, and doubt will lead to alternative thinking, and sometimes that will yield a new and more complete theory.

Thomas Kuhn (Fig. 1.6) elevated this idea to a principle by arguing that science proceeds in a 'normal phase', where researchers work within an accepted framework, filling in details and extending applicability. These periods of steady growth are occasionally interrupted by *paradigm shifts* when old ideas must be abandoned (this is point (iv) above). Kuhn went further by arguing that old and new paradigms must be fundamentally incompatible. (I think that the distinctions

Figure 1.6: Thomas Samuel Kuhn (1922–1996). American physicist, historian, and philosopher of science.

need not be so stark. Some paradigm shifts are subtle and only involve a change in attitude rather that a fundamental upheaval.) Finally, Kuhn maintained that the shift of allegiance from old to new paradigm can only be made as an act of faith. This left some commentators appalled since it undermined the notion that science is objective. But we have already argued that it cannot be objective, and paradigm shifts merely offer another example of the impact of preconceptions and cultural preferences. Indeed, what criteria does one employ to decide between two paradigms? This is the problem of Occam's razor again, with no clear resolution.

1.6   Answerable and Unanswerable in Science

In this section we will apply some of the concepts we have learned to analyze a series of queries that were posed by an eminent scientist 130 years ago. The going will get a little tough because we will delve more deeply into the philosophical underpinnings of modern science. The payoff will be a better understanding of science and the scientific view of reality.

We know that the scientific process is powerful, but what is its scope? Are there questions that science cannot answer? *Why* questions, for example, tend to be problematic in science. Of course we do not seek human motivation as answers to these questions, but causal relationships. For example, the question "Why did it rain yesterday?" might be met with a reply about weather patterns, humidity content of the air, and so on. But the question "Why do people exist?" is more difficult to answer (a possible answer is given below). Certainly the question "Why does the universe exist?" is even more problematic – although even this question may be answerable in the future.

Alternatively, a question like "Why are there eight planets?" is regarded as unanswerable because the solar system is thought to have formed in a largely random maelstrom of self-gravitating gases. The query is regarded as unanswerable because it deals with the ***initial conditions*** that gave rise to the solar system. (We will say much more about this below.) The issue of the number of planets was not always thought to be unanswerable. For example, Johannes Kepler (1571–1630) thought that the orbits of planets had something to do with the regular *Platonic* solids, and since there are five such solids, it was natural to expect five (plus one for the sun, which was mapped to a sphere) planets (Fig. 1.7).

Similarly the question "Why do we have five fingers?" was thought to lie outside the realm of science, but now it can (presumably) be explained in terms of evolutionary fitness. The reason for the chirality of biomolecules[10] was thought beyond science, but the discovery that nature differentiates handedness at a fundamental level in 1956 now permits a plausible explanation.

It is instructive to look at a list unanswerable questions that was created by the distinguished physiologist, Emil du Bois-Reymond (1818–1896) in 1882. Du

---

[10]Many molecules found in living organism display a handedness – think of a screw that goes clockwise or counterclockwise. For some reason all chiral biomolecules are left-handed.

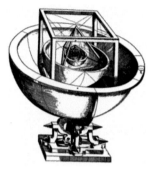

Figure 1.7: Kepler's conception of the solar system consisting of concentric Platonic solids and spheres.

Bois-Reymond coined a phrase to capture the idea of unanswerables: *ignoramus et ignorabimus*, meaning 'we do not know and cannot know'. His ignorabimus short-list consisted of

- the ultimate nature of matter and force
- the origin of motion
- the apparently teleological arrangements of nature
- the origin of life
- the origin of simple sensations
- the origin of intelligence
- the question of free will.

The first three questions concern issues of physics. In the 1880s, the nature of matter was only dimly perceived; Du Bois-Reymond himself referred to atoms as a "useful fiction". These days we know a lot more. Can we say that the first question has been answered? It depends on what du Bois-Reymond meant. We believe that there are four forces and that most (all?) of the fundamental particles of nature have been found. Compared to 1880 this is an enormous advance, and maybe du Bois-Reymond would have been satisfied.

There is a problem, however, because of the word 'ultimate' in du Bois-Reymond's query. How do we know that our particles are the *ultimate particles*? After all, molecules are made of atoms, atoms are made of electrons, protons, and neutrons, and protons and neutrons are made of quarks. Perhaps the story continues in some sort of Escherian infinite regress? So far the international science community has examined nature to a distance of about 10^{-18} m – we do not know what happens at smaller distances. In this sense the question is unanswerable because we never know what might happen around the next length scale corner.

Perhaps, like me, you are unsatisfied with this. With the stress on the word 'ultimate' the question becomes logically impossible to answer. It is completely analogous to attempting to prove that something does not exist. Can you ever be certain that Santa Claus does not exist? Only if you have examined all possible crannies and circumstances of the universe. There are infinitely many equally empty statements in logic and in science; in this sense du Bois-Reymond's first query loses meaning and should not be considered when discussing the boundary of science.

The preceding discussion has glossed an important point. As we have implied, when scientists say that they know the fundamental particles of nature, they mean *so far*, or put better, that they know the particles down to the distance 10^{-18} m. In 1960 the answer was known to 10^{-16} m, and in 1880 it was known to roughly 10^{-5} m. This is a general feature of science. One never says that an object is 1.5 m long but rather that it is 1.5 m measured to an accuracy of, say, 1 mm. This simple and practical expedient avoids all of the philosophical conundrums associated with proving a negative. It is impossible to prove that there are no mice in the cellar, but one *can* say that there are no mice as determined to a certain degree of acuity. This method also avoids a more disturbing philosophical issue: can one ever be certain of a positive? Can you ever say with certainty that you have seen a mouse in the cellar? One can easily construct an infinite regress of mouse properties that the putative mousy creature must possess, and certainty can never be achieved. A scientist avoids the issue by saying 'I have seen a mouse as determined by the following criteria: it was furry, brown, had big ears, . . . '.

In the second query du Bois-Reymond is referring to the *ultimate* source of motion. This was a problem that tormented the Greeks 2500 years earlier. If one traces all the current motion in the world, from a leaf falling, to the planets in their orbits, to the swirl of the galaxies, back in time, whence does it all start? One answer supplied by Parmenides (c. 500 BC) was that the universe has always been, hence there was no beginning, and no possibility to answer the question. Du Bois-Reymond could do no better in 1882. The advent of Einstein's General Relativity and quantum mechanics 40 years later rendered the Greek answer obsolete and has proved to be much more interesting. The current received answer to this question is that the universe began about 14 billion years ago in a quantum fluctuation called the Big Bang, entered a period of superluminal expansion,[11] and then settled into its current state. Motion started with the Big Bang and hence is associated with the *initial conditions* of the universe.

We have run into an important and universal feature of physics: to be predictive, theories need to be able to evolve systems (say of planets or of molecules) into the future. The 'initial conditions' are the starting point and cannot be a prediction of the theory, one can only say what happens subsequently. If one is in a laboratory one can set up or measure the initial conditions and then test the accuracy of the predicted ensuing behaviour. But this is not possible for unique

---

[11]I.e., faster than the speed of light.

systems such as our solar system or the universe itself. Initial conditions for these systems must be inferred from present behaviour and it is impossible to predict what they would be otherwise.

Even knowing initial conditions is sometimes not enough. For example, *chaotic behavior* can make a question unanswerable. Chaos is a general feature of many complicated systems where small errors in the measurement of initial conditions are amplified with time, rendering accurate predictions impossible. This is why weather predictions become less accurate as they go into the future. Chaos was made famous by Edward Lorenz (1917–2008), who colorfully referred to it as the *butterfly effect*. Most complex systems exhibit chaotic behavior, and thus it is impossible to answer precise questions about future outcomes.

And this is not all. The quantum mechanical nature of the universe places fundamental limits on the types of questions that can be asked and the types of answers that can be given. The famous *uncertainty principle*, for example, implies that it is impossible to know both the position and speed of a subatomic particle to arbitrary accuracy. In a similar fashion, it is impossible to predict when a nuclear reaction will occur, only that it will occur in a certain time span with a certain probability.

The third query, 'the apparent teleological arrangements of nature', is not concerned with the boundaries of science, but something more ethereal. Du Bois-Reymond is addressing the apparent element of design in the universe (teleology is the explanation of phenomena by the purpose they serve, rather than their cause). Struggles with teleological arguments date back to antiquity. Democritus (460–370 BC) attempted to construct a mechanistic universe, devoid of the gods. Aristotle rebelled against this idea precisely because it was not anthropomorphic, saying, "Democritus, however, neglecting the final cause, reduces to necessity all the operations of nature.".

For many people the simple explanation is that the universe was indeed designed by God. The response of science to this question has been to dismiss it. One no longer speaks of *why* things are but of *what* things are. In fact, the query is rife with anthropomorphism and is predicated on the prejudice that 'life is different'. Impressed with our ability to think, it is natural to believe that we are different from rocks and water and the rest of nature. But nature does not attempt to impress some of its constituents with the actions of others.

The origin and properties of life are the topic of the last four queries. These days the origin of living matter is not regarded as a great mystery – the early Earth was a cauldron of organic chemicals, and with an abundance of materials, time, and energy, it is perhaps only slightly surprising that amphiphilic[12] bags of self-replicating matter formed. But we are not bugs, and our intelligence does appear truly different. And yet this conceit seems fragile. The past century has seen our exalted position under continuous assault: we are no longer the only

---

[12] An amphiphile is a chemical compound that has a part that is attracted to water and another part that is attracted to fats. Soap is an amphiphile.

tool users, the only communicators, or even the only species with 'civilizations'. And if we are only quantitatively different from an ape, then perhaps we are only quantitatively different from inorganic matter. Perhaps, in the end, the unease over consciousness is simply a failure of imagination rather than a fundamental limitation of physics. Perhaps there is no sacred barrier between life, consciousness, and science.

The final question, the existence of free will, also roiled the ancient philosophers, and remains an open and interesting question today. The main issue is that the laws of physics completely determine the future behavior of everything if (i) the laws are correct in all details, (ii) the laws can be perfectly applied, (iii) the exact initial conditions of everything is known. These conditions are impossible to satisfy in practice, but one can *imagine* that they could be satisfied.

> "We may regard the present state of the universe as the effect of its past and the cause of its future. An intellect which at a certain moment would know all forces that set nature in motion, and all positions of all items of which nature is composed, if this intellect were also vast enough to submit these data to analysis, it would embrace in a single formula the movements of the greatest bodies of the universe and those of the tiniest atom; for such an intellect nothing would be uncertain and the future, just like the past, would be present before its eyes."
>
> — Pierre Simon Laplace, *A Philosophical Essay on Probabilities*, 1814.

Since people are part of nature, one must conclude that a person's actions are as predetermined as those of a molecule of air. This certainly appears to fly in the face of common sense – every day one makes choices that one intuitively feels could have gone the other way. Thus the ideas of *scientific determinism* and *free will* appear to be at odds.[13]

1.7   Why Does Science Exist?

All of this discussion is predicated on the existence of science, a simple fact that has caused some puzzlement in the past. Einstein once quipped, "The most incomprehensible thing about the universe is that it is at all comprehensible." And indeed, one can imagine a universe in which there are no laws at all, or perhaps one in which the laws of physics change rapidly or randomly. It is convenient that we do not live in such a universe since the evolution of sentient life requires the stability of the laws of physics over a time span of billions of years. This ordinary observation has been elevated to the *anthropic principle*, which states that the universe is the way it is because this is what is necessary to permit life. It perhaps

---

[13] We will not venture to solve the problem here.

does not need to be added that many scientists are uncomfortable with this principle – they regard it as untestable and unscientific. All we can say with certainty is that the universe does indeed appear to obey certain mathematical laws and that we seem to have evolved just the level of intelligence needed to discern (some of) them.

Countering this somewhat pessimistic attitude is the observation that if a regularity is discovered, say in the dropping of cannon balls in Pisa in 1590,[14] then the regularity will hold at all places and at all times (these are the two principles mentioned in Sect. 1.1). This is a deep statement about the existence of symmetries in nature, and it is those symmetries that permit a rational science to exist. Thus it is a basic property of nature itself that permits its self-comprehension.

If nature is comprehensible, it is so within the language of mathematics. But why should mathematics be so useful to the physicist? In 1960 the quantum pioneer Eugene Wigner (1902–1995) wrote an influential essay entitled *The Unreasonable Effectiveness of Mathematics in the Natural Sciences* asking this same question. Wigner contended that, "...the enormous usefulness of mathematics in the natural sciences is something bordering on the mysterious and there is no rational explanation for it."

Wigner noted that complicated mathematical concepts, such as complex numbers, find their way into physics surprisingly easily. For Wigner this was part of a trinity of grand mysteries: the existence of laws of nature, the human capability to understand these laws, and the utility of mathematics in formulating the laws.

We have already addressed the first mystery. The second does not seem overly mysterious to me. The mind, after all, has evolved according to the same laws that delineate nature itself. Simple survival implies that it must be acutely attuned to those laws. This is evident in all animal behavior; a cat, for example, employs exquisite knowledge of gravitational dynamics in its movement. And the same cat exhibits higher reasoning, such as object permanence, that is ultimately based on the laws of physics.

Similarly, I would argue that the efficacy of mathematics is not so unreasonable. Mathematics is a logical framework that can be built from a small set of axioms. These axioms were first formulated by Guiseppe Peano (1858–1932) and are of the sort 'zero is a number', 'every number has a successor', 'if $x = y$ then $y = x$'. But whence the axioms? A moment's repose reveals that the axioms arise from the most basic facts of our experience – things like 'either the sun is shining or it is not shining'. But this is *physics* at its most fundamental level! Thus it is no accident that mathematics is relevant to physics.

---

[14]There is good reason to believe that Galileo did not actually conduct this experiment.

REVIEW

Important terminology:

anthropic principle   [pg. 19]

anthropomorphism   [pg. 6]

conservation of energy   [pg. 7]

conservation of momentum   [pg. 7]

induction, deduction   [pg. 2]

Occam's razor   [pg. 4]

scientific determinism   [pg. 18]

teleology   [pg. 17]

theory, model, and law   [pg. 7]

Important concepts:

Kuhn's paradigm shift.

Popper's falsifiability.

Science is a process in which experimental observations are collected and placed in a framework that permits predictions to be made.

The ability to do science rests on symmetries in nature that guarantee that experimental outcomes do not depend on when or where they were obtained.

The application of uncertainty to measured quantities eliminates philosophical problems.

Energy is conserved, momentum is conserved, entropy increases.

Applying Occam's razor is anthropomorphic.

Modeling permits simplification and prediction.

FURTHER READING

J. Henry, *The Scientific Revolution and the Origins of Modern Science*, Palgrave, New York, 2002.

M.C. Jacob, *The Scientific Revolution: a brief history with documents*, Bedford/St. Martin's, New York, 2010.

Samir Okasha, *Philosophy of Science*, Oxford University Press, 2002.

EXERCISES

1. Logic.

   Which of the following are deductive, inductive, IBE, or nonsense.

   (a) Hockey is a sport. Sport is a human activity. Thus hockey is a human activity.

   (b) It is cloudy on rainy days. It is not rainy. Thus it is not cloudy.

   (c) It is cloudy on rainy days. It is not cloudy. Thus it is not raining.

   (d) All mice have green fur. Fluffy is a mouse. Thus Fluffy has green fur.

   (e) All rabbits have green fur. Thus all creatures with green fur are rabbits.

   (f) All rabbits I have seen have green fur. Thus all rabbits have green fur.

   (g) All rabbits I have seen have green fur. A genetic mutation must be causing this.

   (h) All creatures I have seen with green fur have been rabbits. Thus all rabbits have green fur.

2. Laws and Models.

   Which of the following are law, model, or theory.

   (a) the law of supply and demand

   (b) the Standard Model.

3. Evolution.

   In the political debate over evolution one often hears that it is "only a theory". Comment on this claim.

4. Scientific Faith.

   You are asked if a proton in Andromeda has the same mass as one on Earth. You reply that you believe it does but cannot prove it. Your antagonist proclaims this an act of faith and equates it to religious faith, concluding that science is just as faith-based as religion. What is wrong with this argument?

5. Mach on Models.

   Ernst Mach wrote, "In reality, the law [model] always contains less than the fact itself, because it does not reproduce the fact as a whole but only in that aspect of it which is important for us, the rest being intentionally or from necessity omitted." Do you agree with this statement?

6. Multiple Realization.

   Philosophers of science refer to *multiple realizations* as an attempt to explain why subjects such as biology or economics cannot be reduced to physics. The idea is that complex things cannot be precisely defined (i.e., are multiply realized) and therefore a physical explanation of that thing cannot be made. Comment on this idea.

7. Practice with Laws (i).

   I have placed a refrigerator in the middle of a well insulated room and plugged it in. In a fit of pique I leave the door of the fridge open and cool air pours into the room. Does (i) the average temperature in the room rise, (ii) stay the same (iii), or drop?

8. Practice with Laws (ii).

   You have been transported by Darth Maul to the ice planet Hoth where you find yourself in the middle of a large frozen lake. You quickly discover that (i) the ice is completely frictionless (ii) you have your cell phone (iii) you are naked. Point (i) is serious because it means that you cannot get traction when you try to walk – you are stuck in the middle of the lake, doomed to freeze to death. You try calling 911 but there is no reception! How do you escape?

9. Practice with Laws (iii)

   In one of the most famous scenes in movie history, the bad guys try to off Cary Grant by chasing him with a crop duster (the movie is *North by Northwest*). He avoids death by diving in rows of corn and running like heck. Eventually a truck stops on a nearby road and Cary jumps underneath it. The plane slams into the truck and explodes – but the truck doesn't move a whit! What law is violated in this scene?

10. Practice with Laws (iv)

    In the sci-fi classic *Soylent Green*, Edward G. Robinson and Charlton Heston are cops investigating a brutal murder in a dystopian future. Things are so bad that people are reduced to eating wafers produced by the Soylent Corporation (presumably the successor to Exxon-Mobil), and soylent green is the best of the wafers. The climax occurs when Heston realizes that soylent green is made of people. Yum! What law makes this infeasible?

11. Practice with Laws (v)

    Machines have taken over the world! Dude, what if *we're* the computer program! Take these two stock premises, throw in the acting of Keanu Reeves and the direction of the Wachowski brothers and you have the *Matrix* franchise. The kicker is when our heroes realize that the matrix is a virtual reality set up to keep humans happy while they float in giant test tubes. Why bother? To extract human heat and electrical activity as an energy source of course! What law of physics does this flaunt?

12. Unanswerables.

    Classify the following questions as answerable or unanswerable.

    (a) Where do mountains come from?

    (b) How old is the Earth?

    (c) Why do people have five fingers?

    (d) How many planets are there in the Milky Way?

    (e) Why are there that many planets in the Milky Way?

    (f) Why does Jupiter have a mass of $1.89 \cdot 10^{27}$ kg?

13. Science.

    Come up with a theory for why mammals have five fingers that is

    (a) not scientific

    (b) scientific.

    Suggest a test for your scientific theory. How does it relate to Popper's falsifiability criterion?

14. Theory.

    Someone says that scientific theories only make true statements about nature. How do you respond?

15. Scientific Statements.

    Are the following statements scientific?

    (a) Omega-3 aids brain health.

    (b) Fossils were created to look old, but are actually new.

16. Mass.

    Textbooks often define mass as "a quantity of matter". Discuss the utility of this definition.

Doing Science

"It is alleged to be found true by proof, that by the taking of Tobacco, divers[e] and very many do find themselves cured of divers[e] diseases; as on the other part, no man ever received harm thereby. In this argument there is first a great mistaking, and next a monstrous absurdity: . . .when a sick man has his disease at the height, he hath at that instant taken Tobacco, and afterward his disease taking the natural course of declining and consequently the patient of recovering his health, O then the Tobacco forsooth, was the worker of that miracle."

— King James I, *A Counterblaste to Tobacco.*

The previous chapter was fairly theoretical – it's time to get our hands dirty and think about *doing* science. In practice, this means doing experiments; either to explore new phenomena, or more often, to decide something. We define:

An *experiment* is a controlled, reproducible examination of nature intended to arbitrate between competing hypotheses.

In spite of these intentions, an experiment can never settle an issue, it only adds to the evidence supporting (or refuting) a theory or model (reread Chap. 1 if this sounds odd). Of course, often the evidence becomes overwhelming and it becomes foolish to think that experiment has not proven a specific hypothesis (for example, that the Earth is round or that atoms exist).

The majority of a person's contact with science is through medicine. For that reason, this chapter will concentrate on medical science and its peculiar experimental methodology.

© Springer International Publishing Switzerland 2016
E.S. Swanson, *Science and Society*,
DOI 10.1007/978-3-319-21987-5_2

2.1   The Beginnings of Science

The notion that experimentation is a tool to understand nature is relatively new. Many of the ancient Greek philosophers, for example, distrusted earthly, fallible senses.

> "Did you ever reach [truth] with any bodily sense? and I speak not of these alone, but of absolute greatness, and health, and strength, and, in short, of the reality or true nature of everything."
>
> — Plato, *Phaedra*

It was the *Scientific Revolution* (1543–1727) that brought about a lasting change of attitude. An English lawyer, court intriguer, co-founder of the colonies of Virginia and Newfoundland, and member of parliament named Sir Francis Bacon (Fig. 2.1) was instrumental in bringing about this change. Although Bacon had a classical education, he became disenchanted with the diversity of ancient opinion and its search for teleological ultimate cause.

Figure 2.1: Sir Francis Bacon (1561–1626), 1st Viscount of St. Alban. English philosopher, statesman, and essayist.

> "For to what purpose are these brain-creations and idle displays of power .. All these invented systems of the universe, each according to his own fancy [are] like so many arguments of plays ... every one philosophises out of the cells of his own imagination, as out of Plato's cave."

For Bacon the way out of the philosophical morass was clear: empty philosophizing must be replaced by ruthless *empiricism*.[1] Natural philosophers should

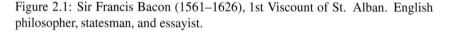

[1] Empiricism is the idea that all knowledge is derived from the senses.

"cut nature to the quick" – only in this way could truth be obtained. Men should not seek final cause, said Bacon, rather they should be satisfied with what is knowable. He also stressed that science could be put to the use of the state, which of course has been taken up with enthusiasm by many modern governments. The reason is doubtlessly related to Bacon's most famous aphorism, "knowledge is power".

Bacon expressed other decidedly modern concepts, stressing the importance of the corpuscular theory,[2] for example, in understanding heat; care in experimentation, insisting that experimenters in different fields should communicate; and that the results of experiments should be meticulously recorded. In spite of these modern ideas, Bacon, like all men, was molded by his time and experiences: it appears that he imagined a form of experimentation that was modelled after legal proceedings.

The new methods of science soon made their way into medicine. One of the first studies made was by naval officer James Lind (1716–1794), who sought to alleviate the suffering of sailors due to scurvy. Lind thought that scurvy was caused by "putrefaction of the body" and that this could be remedied with acidic food. He tested his idea by giving six groups of sailors with scurvy identical diets but with differing supplements (barley water, cider, oranges, etc.) and remarked that those given citrus soon recovered.

In 1775, Sir Percivall Pott (1714–1788) found an association between scrotal cancer in chimney sweeps and exposure to soot, thereby demonstrating a link between occupation and cancer and the existence of environmental *carcinogens*.[3]

Another watershed moment in the development of *epidemiology*[4] came during an outbreak of cholera in London in 1854. John Snow, a local doctor, was skeptical of the miasma theory of disease and sought another explanation for the outbreak. By tracing the addresses of the sick he was able to identify a public water pump as the source of the disease. Later he used statistical methods to show there was a correlation between cholera incidence and water quality. Snow went on to become a founding member of the Epidemiological Society of London.

Finally, we consider the study of childbed fever conducted by the Hungarian physician, Ignaz Semmelweis (1818–1865). At the time, women contracted childbed fever with alarming regularity in European maternity clinics. Semmelweis noted that women attended by physicians tended to contract the disease more often than those attended by midwives. The death of a friend due to infection led him to guess it was physician contact with cadavers that caused childbed fever. Semmelweis tested his idea by requiring his doctors to wash their hands before treating women. The result was an immediate and dramatic drop in the infection rate. Later Semmelweis showed that the opening of a nearby pathological

---

[2]A corpuscle is a particle, so this refers to the idea that matter is made of atoms.
[3]A carcinogen is something that causes cancer.
[4]Epidemiology is the study of causes and effects of disease and other factors that impact health.

anatomy clinic was accompanied by an increase in fever rates – thereby establishing a correlation between the handling of corpses with the incidence of fever.

As bacteria were unknown at the time, Semmelweiss could not explain his findings. His results were ignored or derided by the medical establishment and he lost his job. Semmelweis did not deal with the rejection well and was eventually committed to an asylum, where he died of blood poisoning, probably as a result of being severely beaten by guards.

Important studies continue to be conducted. In the past 60 years it has been established that smoking is linked with lung cancer and that diet is associated with heart disease. These days science is big business. The United States government spends about $143 billion per year on research and development (about ½ of this is on military applications). It is estimated that the government and business spend $100 billion on medical studies every year, while world wide expenditure on biomedical research is around $270 billion per year. These budgets can be compared to the US Department of Energy outlay on particle physics of $¾ billion per year.

2.2 Studies

A *clinical trial* or *study* is an experiment that is typically performed on living things (like people). Because of this, studies have a different complexion than experiments in the physical sciences: people are not as reproducible as crystals or lasers, and the outcomes of experiments are often random (for example, smoking does not always cause cancer). Thus studies tend to seek average effects, require large numbers of subjects, and are prone to human biases.

The competing hypotheses that experiments and studies examine are usually of the type *A causes B* and *A does not cause B*. In the lingo, *A* is an *exposure* and *B* is an *outcome*. Thus a typical study seeks to determine if there is a *causal link* between an exposure and an outcome. For example, one might hypothesize that smoking causes lung cancer in some way; here smoking is the exposure and having cancer is the outcome.

In spite of these goals, studies cannot determine whether an outcome is *caused* by an exposure – they can only measure the *correlation* between the exposure and the outcome. For example noting that smokers tend to have more lung cancer does not (necessarily) mean that smoking causes cancer. Perhaps the cancer is caused by something else that is in turn correlated with smoking.

Correlation does not imply causation.

This difficulty is captured in the phrase, "correlation does not imply causation". Consider a study that finds a correlation between hot dog sales and drowning deaths. One might conclude that swimmers like hot dogs and when they eat too many they tend to get cramps and drown. But a simpler explanation involves no causation at all: hot dog sales go up in the summer, which is when swimming happens. Similarly, one might examine divorce rates over the past 50 years and find a correlation with household television ownership. Maybe watching too much tv leads to divorce, but the correlation is probably due to an increase in both tv ownership and the divorce rate over the past few decades, presumably for independent reasons. A similar observation can be made about the sales of organic food and the rate of autism in the last decade (Fig. 2.2). Although the numbers track together remarkably well, it seems unlikely that this correlation is causal.

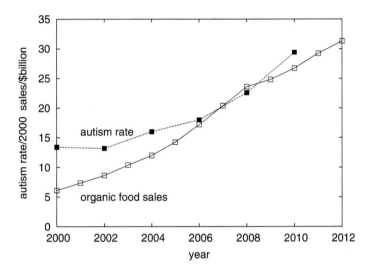

Figure 2.2: Autism rate in 8 year olds (per 2000) and sales of organic foods.

If experiment can never prove causality, how is one to make progress? The answer is that eventually so many experiments find tight correlation between exposure and outcome that it becomes nonsensical to assume a noncausal relationship between them (think of Occam's razor). In an effort to sharpen this, the British epidemiologist Sir Austin Bradford Hill (1897–1991) established a set of conditions necessary to argue for causality.

Bradford Hill Criteria

Analogy Factors similar to a suspected causal agent should be investigated as possible causes.

Biological Gradient There should be a relationship between dose and patient response.

Coherence Controlled laboratory results should agree with epidemiological experience.

Consistency Results of studies should be consistent across a range of factors, such as who conducts the study, when it is conducted, or dosages employed.

Direct Evidence Direct controlled experimental demonstration of an outcome under the influence of an exposure is strong indication of causality.

Plausibility The apparent cause and effect must make sense in the light of current theories. If a causal relationship appears to be outside of current science then significant further testing must be done.

Specificity A specific relationship between outcome and exposure increases the likelihood of a causal relationship.

Strength and Association Strong correlation between outcome and exposure increases the likelihood of a causal relationship.

Temporality The outcome should occur after the exposure.

It is evident that the more of these common sense criteria that hold, the more likely it is that a true causal relationship has been established between exposure and outcome.

2.3 Study Design

2.3.1 Types of Study

Studies can vary widely in their design, depending on goals, financing, and other constraints. A fundamental distinction is between *observational* and *randomized* studies.

observational a study in which the investigator observes rather than influences exposure and disease among participants. Case-control and cohort studies are observational studies.

    case-control an observational study that enrolls one group of persons with a certain disease, chronic condition, or type of injury (case-patients) and a group of persons without the health problem (control subjects) and compares differences in exposures, behaviors, and other characteristics to identify and quantify associations, test hypotheses, and identify causes.

cohort an observational study in which enrollment is based on status of exposure to a certain factor or membership in a certain group. Populations are followed, and disease, death, or other health-related outcomes are documented and compared. Cohort studies can be either prospective or retrospective.

The terms *prospective* and *retrospective* refer to whether trial subjects have or do not have the health outcome of interest at the beginning of a trial.

prospective a study in which participants are enrolled before the health outcome of interest has occurred.

retrospective a study in which participants are enrolled after the health outcome of interest has occurred. Case-control studies are inherently retrospective.

Ex. To test whether power lines cause cancer, researchers questioned cancer patients about how close they lived to power lines.

Ex. To test whether power lines cause cancer, researchers followed the health history of 1000 people living near power lines.

The first example is a case-control study, while the second is a prospective cohort study.

The other study category mentioned was "randomized", which refers to *randomized controlled studies*. These are defined as

randomized controlled a study in which subjects are randomly placed in a two or more groups that receive different treatments,

and are regarded as the most reliable form of study.[5] The term *control* refers to a group that does not receive the test treatment and serves as a basis of comparison for the treatment group. The control group can receive treatment from a known drug or can be given a *placebo*.

A placebo is a sham medicine (often a sugar pill) that is used to minimize the difference between the control and test groups. Specifically, there is a powerful psychological effect associated with the belief that one is being treated. It is therefore important that the control and the test groups both believe they are receiving treatment.

2.3.2 Bias

Researchers give placebos to control groups because they are trying to eliminate *bias* from their studies. There are many sources of bias in studies and researchers go to great efforts to eliminate or reduce them. For example, randomly assigning

---

[5]Many other types of studies exist, including cross-sectional (a survey), screening, and diagnostic.

persons to control and test groups removes **selection bias**. To see the importance of this, consider a researcher with an interest in proving the efficacy of a new drug who selects members of the test group in a study. He could be tempted (unconsciously or otherwise) to choose healthier persons to be in the test group, thereby skewing the results of the study.

In a similar way, a researcher who assesses health outcomes during a trial could be tempted to be generous with people who he knows are in the test group. This is called **interviewer bias**. There is a simple method to control for assessment bias called **blinding**. A blinded study eliminates this bias by not revealing group membership to researchers. It can even be important to blind the study subjects so that they do not know if they are receiving treatment or a placebo. Such studies are called **double blind**.

There are dozens of kinds of bias that can confound the most well intentioned research. Some of these are listed here.

Study Biases

pre-study

> **study design** clear goals and criteria must be decided before a study is undertaken.
>
> **selection bias** patients are not randomized or are not selected according to clear pre-set criteria.
>
> **channeling bias** patients are not added to cohorts with clear pre-set criteria.

in-study

> **interviewer bias** researcher interaction with subjects should be standardized and the researcher should be blinded to the exposure status of the subject.
>
> **recall bias** patients who are asked to recount experience or results can introduce bias. It is preferable to find impartial methods to rate results.
>
> **transfer bias** sometimes studies must follow-up with patients to obtain study results. A policy to deal with patients who cannot be found must be established before the study is made.
>
> **dropout bias** people who leave studies can introduce bias if there is a common reason for withdrawal (such as being too sick to carry on).
>
> **performance bias** studies that depend on procedures (such as surgery) can introduce bias due to time-dependence of ability (for example the surgeon gains experience, or the surgeon has a bad day).

exposure misclassification unclearly defined exposures can introduce bias.

outcome misclassification unclearly defined outcomes can introduce bias.

post-study

> **citation bias** researchers who choose to refer to certain publications, but ignore others, introduce citation bias in their studies.

> **salami slicing** researchers take the results from one study and slice the results into several reports without making clear that the reports are not independent. In this way a single positive trial can appear as many positive trials, giving a false impression.

> **publication bias** researchers who decide not to publish results can skew the public record, leading to bias.

Let's consider examples of these biases.

> Ex. one common example is the perceived association between autism and the MMR vaccine. This vaccine is given to children during a prominent period of language and social development. As a result, parents of children with autism are more likely to recall immunization administration during this developmental regression, and a causal relationship may be perceived.

> Ex. in research on the effectiveness of batterers treatment programs, some researchers use conflictual couples seeking marriage counseling, and exclude court referred batterers, batterers with co-existing mental disorders, batterers who are severely violent, and batterers who are substance abusers . . . and then conduct the research in suburban university settings.

> Ex. using psychology students in studies.

> Ex. the Interphone study on cancer and cell phones determined usage by asking participants to estimate how many hours they used their phones per week.

> Ex. paying subjects (procedural bias).

> Ex. most medical studies have been done on white or black men (sampling bias).

Although these are just some of the ways that studies can be skewed, it is still a daunting list and illustrates just how careful an assiduous researcher must be. It is also the responsibility of readers to be aware of the limitations of any studies they are considering. To assist, a simple test called the *Jadad scale* has been devised by Alejandro Jadad Bechara (1963–) to assess the reliability of studies. Each affirmative answer earns one point: good studies should score 4 or 5, whereas studies scoring 0, 1, or 2 should not be relied on in forming opinion or courses of action.

The Jadad Scale

1. Is the study randomized?

2. Is the study double blind?

3. Were dropouts and withdrawals described?

4. Was the method of randomization described?

5. Was the method of blinding described?

Not all problems are associated with bias; simple methodology can lead to issues in interpreting studies.

> Ex. Members of the same research group went on to publish a comprehensive survey of the content and quality of randomized trials relevant to the treatment of schizophrenia in general. They looked at 2,000 trials and were disappointed in what they found. Over the years, drugs have certainly improved the prospects for people with schizophrenia in some respects. For example, most patients can now live at home or in the community. Yet, even in the 1990s (and still today), most drugs were tested on patients in hospital, so their relevance to outpatient treatment is uncertain. On top of that, the inconsistent way in which outcomes of treatment were assessed was astonishing. The researchers discovered that over 600 treatments – mainly drugs but also psychotherapy, for example – were tested in the trials, yet 640 different scales were used to rate the results and 369 of these were used only once. Comparing outcomes of different trials was therefore severely hampered and the results were virtually uninterpretable by doctors or patients. Among a catalogue of other problems, the researchers identified many studies that were too small or short term to give useful results. And new drug treatments were often compared with inappropriately large doses of a drug that was well known for its side-effects, even when better tolerated treatments were available –an obviously unfair test.
>
> I. Evans *et al.*, *Testing Treatments*.

2.4 Statistics and Studies

People are complex physical systems and do not respond in identical ways to external factors. Thus studies necessarily have an element of randomness to them and conclusions can only be expressed in terms of probabilities. For example, if 100 people are given an experimental drug, 20 may respond well, 30 may experience some benefits, 40 may remain indifferent, and 10 may have serious side effects. Doing the study again will yield different numbers. How is one to interpret the study data?

2.4.1 The Normal Distribution

The way to deal with randomness is with statistics. This can be an intimidating subject, so we are fortunate that a lot can be understood fairly simply.

You are probably familiar with the most basic and famous statistical quantity called the Gaussian or **normal distribution**.

If one were to make a histogram of heights or IQ scores or weights of the American population they would look like normal distributions (Fig. 2.3). The average height for men is about 70 inches, which coincides with the peak of the distribution. The average is called the **mean** and is denoted μ. The shape of the normal distribution (how narrow or fat it is) is given by another quantity called the **standard deviation**, denoted σ. The standard deviation for height is about 3 inches.

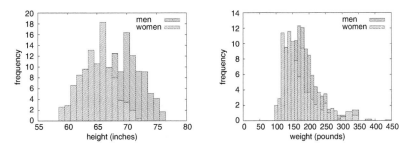

Figure 2.3: Height and weight distributions, US men and women, ages 20–29.
Source: CDC NHANES survey.

These experimental distributions are well approximated by the mathematical normal distribution,[6] shown in Fig. 2.4. The figure illustrates how the standard deviation is related to the shape of the curve. By definition the fraction of the curve between the mean (μ) and the mean plus one standard deviation ($\mu + \sigma$) is 34.1 %. An additional 13.6 % is picked up between $\mu + \sigma$ and $\mu + 2\sigma$, and 2.1 % between 2σ and 3σ. A final 0.1 % remains above $\mu + 3\sigma$.

---

[6]The formula for the normal distribution is $N(x) = \exp[-(x-\mu)^2/(2\sigma^2)]/(\sigma\sqrt{2\pi})$.

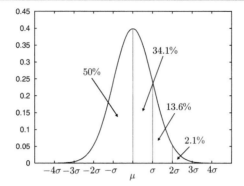

Figure 2.4: The normal distribution.

Standard deviations	Cumulative fraction (%)
μ	50.0
$\mu + \sigma$	84.1
$\mu + 2\sigma$	97.7
$\mu + 3\sigma$	99.9
$\mu + 4\sigma$	$1 - 3.1 \cdot 10^{-5}$
$\mu + 5\sigma$	$1 - 2.9 \cdot 10^{-7}$
$\mu + 6\sigma$	$1 - 9.9 \cdot 10^{-10}$

Ex. The probability of landing within one standard deviation of the mean is 34.1% + 34.1% = 64.2%.

Ex. The mean for IQ tests is defined to be 100 and the standard deviation is about 15. If you have an IQ of 130, 97.7% people have a score lower than you. This is because a score of 130 is 2σ above the mean and the area under the normal curve up to 2σ is 50% + 34.1% + 13.6%.

The normal distribution seems to be everywhere and there is a good reason for this, called the **central limit theorem**. The theorem states that the sum of many random numbers, no matter how they are distributed, approaches a normal distribution (Fig. 2.5). The theorem provides a clue about the pervasiveness of the normal distribution: it must be that height, IQ, etc. are net attributes due to many genetic and environmental factors, each one of them random.

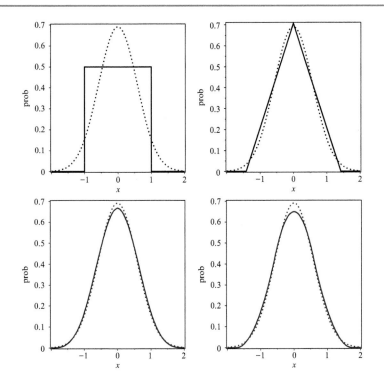

Figure 2.5: The central limit theorem. Clockwise from *top left*: the sum of 1, 2, 3, and 4 random variables.

2.4.2 Error Bars

A main use of the normal distribution is in estimating the reliability of conclusions. Say, for example, that you wish to determine how many Americans are in favor of capital punishment. In principle you could ask everyone in the country, but this would be expensive, and probably impossible to arrange. In practice, pollsters ask a random group of people, called a *sample*. Say 1000 people are asked and the responses are 58 % in favor, 37 % not in favor, and 5 % have no opinion. How reliable is this result?

If you had a lot of money and time, you could ask another 1000 people and check. If you did this 47 times you could make a histogram of the results (Fig. 2.6). As usual, the histogram looks like a normal distribution and we can ask what the mean and standard deviation of the data is. The curve that best reproduces the survey results is shown as a dashed line and tells us that the mean is 59.2 % and the standard deviation (σ) is 0.9 %. This means that our best estimate is that 59.2 % of people agree with the survey question. Also, if we repeat the survey many times, we will find a result between 58.3 % and 60.1 % (59.2 % $\pm$ 0.9 %)

68.4 % of the time. People often say, the average is 59.2 % with an *error bar* of 0.9 % or the average is 59.2 % with 68 % *confidence interval* of 0.9 %.

There is no reason to run 47 different opinion surveys, we could simply make one survey of 47,000 different people and obtain the same agreement rate of 59.2 %. Then we can divide the results into 47 groups, recalculate the averages and make the histogram of Fig. 2.6. In fact this is the way statisticians obtain the confidence intervals that are reported in the media and in studies.

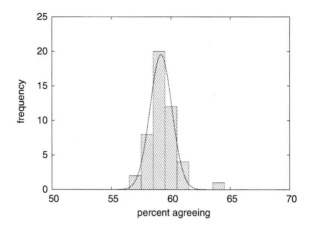

Figure 2.6: Results from your opinion survey.

> Ex. "The AP-GfK poll was conducted March 20–24, 2014. It involved online interviews with 1,012 adults and has a margin of sampling error of plus or minus 3.4 percentage points for all respondents."

What if your bosses wanted *really* accurate results? You might guess that the more people you ask, the more accurate the results. This is correct. The mathematics says that if the number of survey respondents is N then the error bar goes like some number divided by $\sqrt{N}$. This means that if the error with $N = 15$ is $\sigma = 45$ then the error with $N = 1500$ is $\sigma = 4.5$ (an increase of a factor of 100 in survey size means a decrease in the error of a factor of 10).

Data need error bars.

It must be stressed that an experimental number without an error bar is meaningless. What do you care if a survey reports 80 % of people agree with you if you cannot estimate how reliable that number is? A result of 80 % with an error bar of 4 % is much more significant than one with an error bar of 60 %. As we have seen, the way to achieve this is to have a large *sample size*, N. As a rough rule of thumb $N = 1000$ corresponds to error bars of a few percent, while samples with 10s of subjects give errors of tens of percent.

> Ex. A survey of 400 people reveals that 32% prefer vanilla ice cream. The 68% confidence interval is 24%–40%. How many people would have to be interviewed to obtain a confidence interval of 28%–36%?

> A. The old error bar was 8%; the new one should be 4%, which means the sample size should go up by a factor of 4, so 1600 people are required.

While a sufficiently large sample is required to draw reliable conclusions, a researcher (and people who consult studies) must also pay attention to the quality of the sample. For example a survey of the popularity of dub step among 1000 college students would yield quite different results from a survey of 1000 retired people. This is an example of selection bias.

> Ex. Your grandmother and uncle both developed bunions while wearing Converse sneakers. You do *not* conclude that wearing Converses causes bunions because you realize that this is a survey of size $N = 2$ with strong selection bias.

Anecdote is not data.

2.4.3   Hypothesis Testing

The concepts of standard deviation and confidence intervals are central to **hypothesis testing**. Recall that an experiment is an attempt to arbitrate between competing hypotheses. It is traditional to call one of these hypotheses the **null hypothesis**. A null hypothesis is usually a statement that the thing being studied produces no effect or makes no difference. An example is "This diet has no effect on people's weight." Normally an experimenter frames a null hypothesis with the intent of rejecting it: that is, he seeks to show that the thing under study *does* make a difference.

Possible experimental outcomes are traditionally represented in a 2×2 grid as shown in Table 2.1.

The check marks indicate correct conclusions – either rejecting a false statement or accepting a true one.[7] A type I error is also called a false positive and means that a false statement has been accepted as true. A type II error implies that a true statement has been rejected as being false.

---

[7]The entry "accept null hypothesis" should more properly be called "fail to reject the null hypothesis". We stick with the first because it is less wordy.

Table 2.1: Experimental outcomes.

	Null hypothesis true (no effect)	Null hypothesis false (effect exists)
Reject null hypothesis	Type I error (false positive)	✓
Accept null hypothesis	✓	Type II error (false negative)

Ex. Type I error: a fire alarm goes off, yet there is no fire.

Ex. Type II error: you are pregnant but your test does not turn blue.

The probability of a type I error is called alpha (α), while the probability for a type II error is called beta (β). Clearly one wants to minimize these probabilities so that the chance of making a correct conclusion is maximized. An example will illustrate the importance of these numbers.

Ex. About 1 in 10000 people have hepatitis C. An accurate test promises a false positive rate of 1.5%. What do you tell your cousin who tests positive? The actual probability for having the disease is 0.01%, which means that the odds of the test giving a false result is 150 times higher than the odds of actually having hepatitis C. Your cousin should not worry, although seeking another test is advisable.

Power is defined as the probability of rejecting the null hypothesis given that the null hypothesis is false. Since beta is the probability of accepting a null hypothesis given that it is false, we derive

$$\text{power} = 1 - \beta. \tag{2.1}$$

It is desirable to have power as close as possible to 1.0 (values around 0.8 are a typical goal). Power depends on two quantities the researcher cannot control: the size of the effect being measured and the standard deviation of the sample data. In general, the larger the effect and the lower the standard deviation, the higher the power. It also depends on two quantities the researcher can control: the sample size and the desired statistical significance of the study. The higher the sample size and the lower the statistical significance, the higher the power.

Ex. With a power of 0.7, if 10 true hypotheses are examined 3 will be incorrectly rejected.

We have introduced the idea of **statistical significance**. Informally, this is a measure of the chance of obtaining a given effect. It can be defined formally as a *p-value*, which is the probability of getting the result you did (or a more extreme result) given that the null hypothesis is true. This is also known as a **significance criterion**. It is desirable to have a low p-value so that one can be reasonably sure that the null hypothesis should be rejected. In practice, the researcher selects a desired value for alpha (recall this is the probability of accepting the null hypothesis if it false), computes the p-value, and rejects the null hypothesis if the p-value is less than alpha. A typical value for alpha is 0.05.

Unfortunately the interpretation of a p-value is regularly mangled in the media and by scientists themselves. Let us state clearly:

> There is no simple relationship between a p-value and the probability of a hypothesis.

Specifically, the p-value is a probability of observing something given a hypothesis (i.e., the null hypothesis). You are *not allowed* to turn it around and say that it is related to the probability of a hypothesis given your observation. To illustrate, consider the statements (a) the probability of being a woman given that you are in the House of Representatives[8] is 18 % (b) the probability of being in the House of Representatives given that you are a woman is 18 %. Clearly (a) makes sense while (b) is nonsense[9]

Making statements about beliefs in hypotheses is not the only way things can go wrong. A list of common mistakes includes:[10]

Mistakes with p-value

1. The p-value is not the probability that the null hypothesis is true.

2. The p-value is not the probability that a finding is a fluke (this error is very common).

3. The p-value is not the probability of falsely rejecting the null hypothesis.

4. The p-value is not the probability that a replicating experiment would not yield the same conclusion.

5. The (1-p)-value is not the probability of the alternative hypothesis being true.

---

[8] As of the 113th Congress there are 79 women out of 435 representatives.

[9] There is a way to calculate (b) called Bayes' Theorem.

[10] Source: M.J. Schervish, *P Values: What They Are and What They Are Not*, The American Statistician **50**, 203–206 (1996).

6. The significance level of the test is not determined by the p-value.

7. The p-value does not indicate the size or importance of the observed effect.

It is common for a study to seek to find a statistically significant difference between two outcomes. For example, drug A may be more effective than drug B, or women may score higher in IQ tests than men. Some mathematics that we need not go into allow us to apply the ideas of this section to this situation.

Assume that we have two data sets, one with average $\bar{A}$ and standard deviation σ_A, and the other with average $\bar{B}$ and standard deviation σ_B. We want to know how likely it is that the actual (as opposed to the measured) averages are equal. If these values are $\bar{A} = 1$, $\sigma_A = 0.1$ and $\bar{B} = 2.8$, $\sigma_B = 0.03$, then it is very unlikely that the actual averages are equal since the measured averages differ by 1.8 and the errors are quite small (Fig. 2.7 left). Alternatively, if $\sigma_A = 0.9$ and $\sigma_B = 2.2$ then it is much more likely that the actual averages are in fact equal (Fig. 2.7 right).

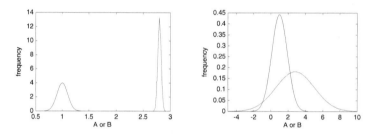

Figure 2.7: Two measured distributions.

The situation can be quantified with the following result: the probability distribution of the sum of two normal distributions is another normal distribution with mean $\bar{A} + \bar{B}$ and standard deviation $\sqrt{\sigma_A^2 + \sigma_B^2}$.[11]

Since this result applies to the difference of two normal distributions as well as to the sum, we can use it to quantify our example problems. In the first case the difference of the means is 1.8 while the new standard deviation is $\sigma = \sqrt{0.1^2 + 0.03^2} = 0.104$. In the second case the difference is still 1.8 but the standard deviation is $\sigma = \sqrt{0.9^2 + 2.2^2} = 2.38$. In the first case 1.8 is 17.3σ removed from zero, while in the second it is 0.76σ removed. Thus the probability that the first difference is consistent with zero is tiny (about 10^{-64}), while the second is about 44 %.

---

[11]This result actually applies for the exact means and standard deviations. We will use it with the measured means and standard deviations, which are the best approximations we have to the (unknown) actual values.

If one assumes a null hypothesis that the two actual averages are identical, then the probability of obtaining a difference of means greater than 1.8 is 10^{-64} in the first case and 44 % in the second. Thus the p-value for the first experiment is tiny and for the second experiment is 0.44.

Assigning probabilities like this is called a ***hypothesis test***. More general hypothesis testing is possible. Consider, for example, the data in Fig. 2.8. The data are live birth sex odds (boys/girls) in the Russian Federation. The authors of the study from which these graphs were constructed hypothesized that the Chernobyl nuclear disaster in 1986 is the cause of the increase in the ratio seen in the figures. As proof they offered the right hand figure which shows a jump in the ratio in 1986. Their hypothesis is a step function like

$$\text{sex odds} = \begin{cases} a & \text{if year} < y \\ b & \text{if year} > y \end{cases}, \tag{2.2}$$

which is a three parameter model (a, b, y) of the data. A fit yields $y = 1986.9 \pm 0.4$, which agrees with their contention about Chernobyl.

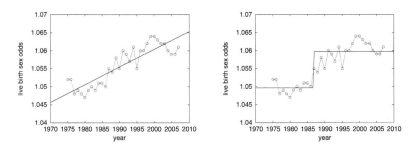

Figure 2.8: Live birth sex odds in the Russian Federation with two model fits.
Source: H. Scherb and K. Voigt, Environmental Science and Pollution Research, **18**, 697–707 (2011).

But there are problems: (i) the data have no error bars so it is difficult to see if the rise in sex odds is due to chance, (ii) even if there is a jump in 1986, there is no reason to associate this with Chernobyl; after all this was the year that *Walk like an Egyptian* was released by the Bangles, (iii) one should compare the model to alternative models to assess how reliable it is. To illustrate the last point I fitted the data to a straight line (left figure). The resulting quality of fit is nearly identical to the step function, and is achieved with one fewer adjustable parameter. Although there is an interesting question about the change seen in the sex odds, one cannot conclude that the Chernobyl disaster has anything to do with it.

Figure 2.9 illustrates another way in which data analysis can go wrong. In this case, the data have error bars and are fit nicely by a straight line. But something is wrong. Recall that an error bar usually represents the one-σ variation in the data. Thus if the data really do follow the line indicated in the figure, 32 % of the data points should lie more than an error bar away from the line. But *all* of the data lie closer (substantially closer) to the line than their error bars warrant.

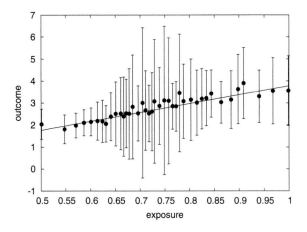

Figure 2.9: Outcome vs. exposure in a study.

2.4.4   The Statistics of Bias

In the past two sections we have examined statistical methods as applied to studies and a rather long list of the ways studies can go wrong. In this section we will use statistics to gain a quantitative understanding of how studies fail.

The simplest statistical concern is a sample size that is too small. The sample size refers to the number of subjects in the study – too few subjects means that conclusions are not statistically significant (we have already seen this in the discussion of the p-value). This is a serious, but common, problem, typically brought on by practical constraints, such as lack of funding or time.

The subtlest issue concerns psychological pressures on researchers. Scientists are under institutional and personal pressure to make important discoveries. When this is combined with financial pressure to publish frequently, it is easy for biases to subvert the accuracy of a study.

> "Simulations show that for most study designs and settings, it is more likely for a research claim to be false than true. Moreover, for many current scientific fields, claimed research findings may often be simply accurate measures of the prevailing bias."
>
> — John Ioannides.

John Ioannides (1965–), who studies clinical trial methodology has noted the following general features that carry negative impact on study fidelity[12]:

1. The smaller the studies conducted in a scientific field, the less likely the research findings are to be true.

2. The smaller the effect sizes in a scientific field, the less likely the research findings are to be true.

3. The greater the number and the lesser the selection of tested relationships in a scientific field, the less likely the research findings are to be true.

4. The greater the flexibility in designs, definitions, outcomes, and analytical modes in a scientific field, the less likely the research findings are to be true.

5. The greater the financial and other interests and prejudices in a scientific field, the less likely the research findings are to be true.

6. The hotter a scientific field (with more scientific teams involved), the less likely the research findings are to be true.

Much of what goes wrong is a simple consequence of false positives. To see this, consider a list of 1000 hypotheses that are deemed interesting enough to test. We will assume that 100 of these are actually true. We also assume that alpha = 0.05 and beta = 0.2. Thus when the 1000 studies are completed there will be $0.05 \times 900 = 45$ false positive results and $0.2 \times 100 = 20$ false negative results. This leaves 80 true hypotheses that have been confirmed (the power is 0.8). The net result is that the researcher believes she has $45 + 80 = 125$ true results, but $45/125 = 36\%$ of these are in fact false. Although the false positive probability is relatively small, it still skews the understanding of the scientific landscape because so few of the hypotheses are actually true (or so many are actually false).

Yet another simple statistical effect can confound the interpretation of large trials. Consider a study with alpha set to 0.05, so that the researcher seeks a p-value of 0.05 or less. Assume further that no effect exists, so that the null hypothesis is true. Since the p-value is the probability of obtaining the experimental result if the null hypothesis is true, on average one experiment in 20 will *not* find the result; i.e., will reject the null hypothesis and claim a discovery. This is called the *multiple comparisons problem* or the *look elsewhere effect*.

> If you look often enough you will find something.

Quite simply, if an experiment examines enough different outcomes for a given exposure, it is bound to observe a statistical fluctuation that will be interpreted

---

[12]PLoS Medicine, **2**, e124 (2005).

as a real rejection of the null hypothesis. The physicist Robert Adair (1924 –) has called studies that examine many possible outcomes *hypothesis generating experiments* because they can used to identify outcomes to be tested with subsequent studies. They cannot be used to claim real outcomes because of the look elsewhere effect.

> Ex. A 1992 Swedish study tried to determine whether power lines caused poor health effects. The researchers surveyed everyone living within 300 meters of high-voltage power lines over a 25-year period and looked for statistically significant increases in rates of over 800 (!) ailments. The study found that the incidence of childhood leukemia was four times higher among those that lived closest to the power lines, and it spurred calls to action by the Swedish government. The problem with the conclusion, however, was that they failed to compensate for the look-elsewhere effect; in any collection of 800 random samples, it is likely that at least one will be at least 3 standard deviations above the expected value, by chance alone. Subsequent studies failed to show any links between power lines and childhood leukemia.

A careful researcher can account for the look elsewhere effect. A simple, but approximate, way to do this is to replace the condition for significance, $p < \alpha$ with $p < \alpha/n$, where n is the number of tested outcomes. This will make a dramatic difference in large-scale studies such as the Swedish study.

Notice that the look elsewhere effect applies equally well to a great many studies that consider the same exposure. Thus 800 studies with $\alpha = 0.05$ will find (on average) 40 false results just as surely as one large study. A current popular health concern and subject of many studies is the effect of electromagnetic radiation on health.[13] The look elsewhere effect *requires* that some of these studies will find an effect. Of course, subsequent studies (if they exist) will not find the same health outcome, but will "discover" a different, and random, outcome. How do you think that these findings will be reported in the media?

> Ex. In 2007 the BBC reported that the president of Lakehead University refused to install wifi on campus because he believes that "microwave radiation in the frequency range of wi-fi has been shown to increase permeability of the blood-brain barrier, cause behavioral changes, alter cognitive functions, activate a stress response, interfere with brain waves, cell growth, cell communication, calcium ion balance, etc., and cause single and double strand DNA breaks."

> Ex. A web site devoted to spreading alarm about electromagnetic radiation reports, "dozens of published papers have found links between living near power line electromagnetic radiation and a range of

---

[13]This topic will be revisited in Chap. 5.

> health woes including brain cancer and leukemia, breast cancer, birth
> defects and reproductive problems, decreased libido, fatigue, depres-
> sion, blood diseases, hormonal imbalances, heart disease, sleeping
> disorders, and many others."

Yes, the media report *all* the random and false discoveries as real. Perhaps it goes without saying that when a single exposure can lead to so many deleterious outcomes one should immediately suspect the look elsewhere effect.

The net effect of all of these issues can be studied by examining published results and seeing how often they are confirmed by follow-up studies. The results are not encouraging. When the pharmaceutical firm Amgen attempted to replicate 53 landmark studies it was only able to reproduce 6 of them. Similarly, researchers at Bayer HealthCare were only able to reproduce one quarter of 67 seminal studies.

A less direct way to test study reliability is to check results as a function of who obtains them. The following table summarizes the results of trials testing the efficacy of acupuncture by country. It is not sensible that trials in North America find a favorable effect in 49 % of studies, while those in Asia find a favorable result 100 % of the time. Some difference, presumably cultural and historical, must be leading to significant bias in the Asian or North American trials (Table 2.2).

Table 2.2: Controlled clinical trials of acupuncture by country of research.

Country	Trials	Trials favoring
USA	47	25
Canada	11	3
China	36	36
Taiwan	6	6
Japan	5	5
Hong Kong	3	3

Source: A. Vickers *et al.*, Controlled Clinical Trials **19** 159 (1998).

Finally, studies reveal that 80 % of nonrandomized studies turn out to be wrong, along with 25 % of randomized studies, and 10 % of large scale randomized studies. These are not encouraging numbers and indicate the power of the forces arrayed against producing quality research and the difficulty in teasing results out of extraordinarily complex systems such as the human body.

2.5 Improving Study Reliability

Increasing awareness of the problem of unreliable studies has led to several efforts to improve methodology. One outcome of this effort was the CONSORT statement of minimum requirements for reporting randomized trials that was formulated by an international group of medical journal editors, clinical researchers, and epidemiologists. The statement provides a standard way for authors to prepare reports of trial findings, facilitates complete and transparent reporting, and aids critical appraisal. You can find out more at www.consort-statement.org.

In 2000 the US National Institutes of Health (NIH) instituted a web site called ClinicalTrials.gov for tracking publicly funded clinical studies. The site is a Web-based resource that provides patients, their family members, health care professionals, researchers, and the public with access to information on publicly and privately supported clinical studies on a wide range of diseases and conditions. Of course, it also addresses issues with publication bias and salami slicing.

The CONSORT and NIH web sites are tools meant for professionals. Fortunately, the *Cochrane Collaboration* was created to look after the rest of us. The collaboration is a non-profit global organization of independent health practitioners, researchers, and patient advocates that is dedicated to producing credible, accessible health care information that is free from conflict of interest. The chief product is a series of systematic reviews that address specific health care questions. These reviews are available online at summaries.cochrane.org.

> Ex: Entering "fish oils for the prevention of dementia in the elderly" into the Cochrane Summary search field yields a report with the following statement.
>
> "The results of the available studies show no benefit for cognitive function with omega-3 PUFA supplementation among cognitively healthy older people."

REVIEW

Important terminology:

biases: publication, recall, citation, sampling, procedural   [pg. 31]

blinding   [pg. 32]

control group   [pg. 31]

study types: cohort, randomized, observational, case controlled   [pg. 31]

exposure and outcome   [pg. 28]

mean and standard deviation   [pg. 35]

placebo   [pg. 31]

type I and type II errors   [pg. 40]

alpha: the probability of rejecting the null hypothesis given that it is true. [pg. 40]

beta: the probability of accepting the null hypothesis given that it is false. [pg. 40]

p-value: the probability of getting the results found given that the null hypothesis is true.   [pg. 41]

power: 1 - beta, or the probability of rejecting the null hypothesis given that it is false.   [pg. 40]

Important concepts:

Correlation is not causation.

Anecdote is not data.

The Jadad scale.

The Bradford Hill criteria.

The Central Limit Theorem.

An experiment is a controlled, reproducible examination of nature intended to arbitrate between competing hypotheses.

The normal distribution is common because it is equivalent to the sum of many random variables.

Error bars permit assessing the reliability of conclusions.

There is no simple relationship between a p-value and the probability of a hypothesis.

When many hypotheses are false, truth can be overwhelmed by false positives.

The look elsewhere effect can give rise to false positives.

FURTHER READING

Imogene Evans, Hazel Thornton, Iain Chalmers, and Paul Glasziou, *Testing Treatments*, Pinter and Martin, 2011.

Ben Goldacre, *Bad Science*, Faber and Faber, 2010.

EXERCISES

1. It is common to hear that children at a birthday party are running wild because they are on a "sugar high". Suggest a causal mechanism for this. Suggest a noncausal mechanism. Suggest a way to test both ideas.

2. It is common to hear that someone caught a cold because they were cold. Suggest a causal mechanism for this. Suggest a noncausal mechanism.

3. In an effort to examine whether exposure to electromagnetic radiation is associated with cancer, a study examined cancer rates in power line workers and found significantly higher incidence of skin cancer. Suggest a causal link for this correlation. Suggest a noncausal link.

4. You wish to test the hypothesis that a coin is fair (i.e., the odds of coming up heads is $\frac{1}{2}$). You flip the coin six times and obtain 5 heads.

   (a) Compute the probability of obtaining at least 5 heads.

   (b) Take the null hypothesis to be that the coin is fair. If your test criterion is alpha < 0.05, do you accept or reject the null hypothesis?

   (c) What p-value would you assign to the statement that the coin is fair?

5. Re-read the quotation at the beginning of this chapter. What point is King James making?

6. Suspicious Data.

   Look at Fig. 2.9 again. If the error bars represent a 90 % confidence interval, how many points do you expect (on average) to lie further from the fit line than their error bar?

7. A sample of crime-scene DNA is compared against a database of 10,000 people. A match is found and the accused person is brought to trial where it is stated that the odds of two DNA samples match is 1 in 5000. The prosecutor, judge, and jury all interpret this to means the odds the suspect is guilty is 4999 out of 5000. What do you say?

8. In 2006 the Times of London reported that "Cocaine floods the playground". The story noted that a government school-yard survey found that cocaine use in London schools had risen from 1 % in 2004 to 2 % in 2005, which they reported as "cocaine use doubles". The survey asked school children about their use of dozens of illegal substances. Why didn't government statisticians break this story?

9. To study the dangers of cellphones, researchers question 100 people with brain cancer to determine their rate of cellphone usage. What type of study is this?

10. Patients in a case controlled study on stomach cancer and antacid consumption are asked how many antacids they eat on average per month. What type of bias does this introduce to the study?

11. An experiment claims to find an effect with a p value of 0.04. If the experiment is repeated 100 times, about how many times will the effect not be seen?

12. A December 13, 2011 New York Times article, "Tantalizing Hints but No Direct Proof in Particle Search", reported

> "The Atlas result has a chance of less than one part in 5000 of being due to lucky background noise, which is impressive but far short of the standard for a 'discovery', which requires one in 3.5 million odds of being a random fluctuation."

What is wrong with this statement?

13. Combining Results. A manufacturer says that the length of his widgets is 1.3 m with an error of 1 mm, as measured by a sample of size $N = 10,000$. You buy 1000 widgets and measure an average length of 1.32 m with an error of 3 mm. Do you believe the manufacturer?

14. Consider the 1000 hypotheses scenario of Sect. 2.4.4 again, but this time assume alpha = 0.05 and beta = 0.79 (this corresponds to a power of 0.21, which is typical of neuroscience studies). What fraction of "true" hypotheses are actually true?

15. Height.

    You are 73 inches tall. What percentage of American men are taller than you?

16. Identify the types of studies mentioned in Sect. 2.1.

    (a) Lind's scurvy study

    (b) Snow's cholera study

    (c) Semmelweis's childbed fever study.

17. Mammograms and Cancer.

    Consider the following 2×2 table for a mammograms and cancer

    (a) What is the null hypothesis?

	Cancer (1 %)	No cancer (99 %)
Test pos	80 %	9.6 %
Test neg	20 %	90.4 %

(b) What are alpha and beta?

(c) Assume 1 % of people actually have cancer, what is the probability that you have cancer if you get a positive test result?

18. Weight distribution.

Look again at the distribution of weights in Fig. 2.3. It does not look very much like a normal distribution. Come up with possible reasons for this.

19. Scientists conducting the European Union INTERPHONE study of cancer and cellphone use disagreed about the validity of their study because patients were asked about their typical cellphone usage. What were the researchers worried about? How could the problem be circumvented?

20. Daniel in Babylon.

Read Daniel 1.5–15. What Jadad score do you give Daniel's study?

21. The following quotation is from an article by Jonah Lehrer which was published in the Dec 16, 2011 issue of *Wired*. Comment on his observations in light of what you have read in this chapter.

"When doctors began encountering a surge in patients with lower back pain in the mid-20th century, as I reported for my 2009 book *How We Decide*, they had few explanations. The lower back is an exquisitely complicated area of the body, full of small bones, ligaments, spinal discs, and minor muscles. Then there's the spinal cord itself, a thick cable of nerves that can be easily disturbed. There are so many moving parts in the back that doctors had difficulty figuring out what, exactly, was causing a person's pain. As a result, patients were typically sent home with a prescription for bed rest."

"This treatment plan, though simple, was still extremely effective. Even when nothing was done to the lower back, about 90 percent of people with back pain got better within six weeks. The body healed itself, the inflammation subsided, the nerve relaxed."

"Over the next few decades, this hands-off approach to back pain remained the standard medical treatment. That all changed, however, with the introduction of magnetic resonance imaging in the late 1970s. These diagnostic machines use powerful magnets to generate stunningly detailed images of the body's interior. Within a few years, the MRI machine became a crucial diagnostic tool."

"The view afforded by MRI led to a new causal story: Back pain was the result of abnormalities in the spinal discs, those supple buffers between the vertebrae. The MRIs certainly supplied bleak evidence: Back pain was strongly correlated with seriously degenerated discs, which were in turn thought to cause inflammation of the local nerves. Consequently, doctors began administering epidurals to quiet the pain, and if it persisted they would surgically remove the damaged disc tissue."

"But the vivid images were misleading. It turns out that disc abnormalities are typically not the cause of chronic back pain. The presence of such abnormalities is just as likely to be correlated with the absence of back problems, as a 1994 study published in The New England Journal of Medicine showed. The researchers imaged the spinal regions of 98 people with no back pain. The results were shocking: Two-thirds of normal patients exhibited 'serious problems' like bulging or protruding tissue. In 38 percent of these patients, the MRI revealed multiple damaged discs. Nevertheless, none of these people were in pain. The study concluded that, in most cases, 'the discovery of a bulge or protrusion on an MRI scan in a patient with low back pain may frequently be coincidental.'"

Pseudoscience

*"We inhabit a universe where atoms are made in the centers of stars;
where each second a thousand suns are born; where life is sparked
by sunlight and lightning in the airs and waters of youthful planets;
where the raw material for biological evolution is sometimes made by
the explosion of a star halfway across the Milky Way; where a thing
as beautiful as a galaxy is formed a hundred billion times - a Cosmos of quasars and quarks, snowflakes and fireflies, where there may
be black holes and other universes and extraterrestrial civilizations
whose radio messages are at this moment reaching the Earth. How
pallid by comparison are the pretensions of superstition and pseudoscience; how important it is for us to pursue and understand science,
that characteristically human endeavor."*

— Carl Sagan, *Cosmos*

In the last chapter we have learned how studies are designed, how they are
interpreted, and how they can go wrong. Unfortunately, bad science happens.
But, as we saw in Chap. 1, the scientific method weeds out incorrect results and
progress towards some version of "truth" is made.

Alas, this does not mean that bad science necessarily vanishes – it can live on
as *pseudoscience*. We have seen the term pseudo-science in Chap. 1 where it was
used by Karl Popper to refer to things purporting to be scientific but that were not
falsifiable. Popper's exemplar was Freud's psychoanalytic theory, which could
find an explanation for any behavior.

These days pseudoscience has a somewhat different meaning[1]: it is a claim
that is presented as scientific but that does not follow the scientific method. Here
we will generalize this slightly to any belief about the natural world that has no
supporting evidence or violates well established scientific laws.

One might guess that pseudoscience does not affect much in society, but this
fails to appreciate the scale of the problem. A rough guess is that science and
pseudoscience are equally represented in western culture. This ratio can drop

---

[1] Hence the lack of a dash.

© Springer International Publishing Switzerland 2016
E.S. Swanson, *Science and Society*,
DOI 10.1007/978-3-319-21987-5_3

substantially lower in less well developed parts of the world, where, quite simply, science is less relevant to the average person.

Closer to home, the size of the issue can be judged with the following figures

Pseudoscience (?)	Cost/year in USA ($ billions)
Therapeutic magnets	0.5
Missile defense	10
Power line mitigation	23
Vitamin industry	23
Complementary and alternative medicine	34
Diet industry	46

These wastes of public or personal money could be negligible compared to those that will come in the near future. Questions of managing the Earth's finite resources, humanity's demand on these resources, energy supply, and climate must all be dealt with. It must surely be better to tackle these issues with real scientific methods rather than with the fantasies created in a pseudoscientific parallel universe.

3.1   Pseudoscience and Society

If the outcome of an experiment holds in Brazil it also holds in Germany; if it held 100 years ago, it holds now. These are hallmarks of science of deep significance that have been discussed in Chap. 1.

Because pseudoscience is not self-correcting, it tends to change with time and place. One hundred years ago seances and other attempts to communicate with the dead were the rage. Fifty years ago UFOs, dianetics, orgonomy, and Atlantis were popular.[2] Thirty years ago it was commonplace to hear about ancient evidence of alien visitors, the power of crystals, and biorythms. Some of these foolish things have vanished, and I suspect that you may not have heard of dianetics, orgonomy, Erich von Däniken, or Immanuel Velikovsky.[3]

---

[2] "I think that it is much more likely that the reports of flying saucers are the results of the known irrational characteristics of terrestrial intelligence than of the unknown rational efforts of extra-terrestrial intelligence." – Richard Feynman.

[3] In brief: dianetics was created by science fiction writer L. Ron Hubbard (1911–1986) and is a way to remove "engrams" to achieve at state called "clear". Orgonomy, an invention of Wilhem Reich (1897–1957), is a study of orgone, which is a cosmic life force that is everywhere. Erich von Däniken (1935–) is a Swiss lifelong minor criminal who wrote *Chariots of the Gods*, which promoted the thesis that ancient aliens visited Earth and influenced early culture. Immanuel Velikovsky (1895–1979) wrote *Worlds in Collision* saying that ancient cataclysms caused by encounters with other bodies in the solar system could explain many biblical stories. Biorythmics is the idea that people's physical, emotional, and intellectual states follow fixed periodic cycles starting from birth.

Alternatively, some folklore has remarkable longevity; astrology, homeopathy, and acupuncture among them. Clearly these ideas resonate with people. We shall examine why in Sect. 3.4.2.

Travel is rightly regarded as a way to expand one's perspective. It is all too easy to believe that we have the right way to do things, whereas a bit of travel will reveal that many right ways exist. It can also expose your pseudoscientific beliefs (or those of others) in surprising ways. Consider the following examples of folk wisdom that I have heard (often out of the mouths of scientists) in recent travels:

Brazil: faith healing is common.

Czech Republic: the alcohol made with Czech brewing practices is better for you.

France: homeopathy is highly regarded.

Germany: wearing wet swimming trunks is unhealthy.

Germany: drinking cold beverages in hot weather is unhealthy.

Italy: coffee made with Italian espresso machines is better for you.

Pakistan: Coca Cola is a toxin.

I have never heard of several of these "obvious truths". How truthful can they be? Europeans tend to be far more distrustful of genetically modified food crops than Americans. Who is right?

It is even possible for pseudoscientific beliefs to clash. I recently spent a frustrating hour in a British grocery store looking for a beverage that was not artificially sweetened. Of the dozens of products I checked, *none* were sweetened with natural sugar. It turns out the British believe that natural sugar is rotting the teeth of their young and hence have banished it. Why did I care? Because I had heard that artificial sugar is "bad for you". Are either of these positions based on scientific evidence?

Relative to parts of the world, North America and Europe appear as bastions of scientific rationalism. But pseudoscience still abounds. Consider the results of a recent Gallup poll (Table 3.1). Almost all of these topics violate some principle of common sense or science, and yet belief persists. Now think of the far greater array of things that are not obviously wrong. How many do you believe? Do you have good reasons for your belief?

Ex. How many of the following do you believe?

- climate change is a fabrication
- being physically cold can give you a cold
- taking vitamin C is good for colds
- fluoridation is dangerous
- people can be "left-brained" or "right-brained"

Table 3.1: Gallup Poll of American Beliefs in the Paranormal, 2005.

Topic	Believe (%)
Extrasensory perception	41
Houses can be haunted	37
Ghosts exist	32
Telepathy	31
Clairvoyance	26
Astrology	25
Aliens have visited the earth	24
Communication with the dead	21
Witches	21
Reincarnation	20
Channeling	9

- cell phones can cause cancer
- birds navigate by the Earth's magnetic field
- magnetic bracelets can alleviate arthritis
- travel to the stars is possible
- the MMR vaccine can be dangerous
- eating fish oils helps the brain
- GMO foods are dangerous

It is possible a few of these things will turn out to be true, but the point is that many people believe many things without good reason. It would be unfair to blame the believers since they merely seek knowledge, just like the most austere and rigorous of scientists. What has gone wrong?

3.2   Bad Science, Bad Scientists

3.2.1   Pseudo-origins

If science is created by scientists doing sciency stuff, where does pseudoscience come from? We have already seen a few examples: Hubbard, Reich, von Däniken, and countless others were charlatans who created scientific sounding (at least to nonscientists) theories of reality that were literally bought by many people. All of them became rich from their deceptions. In these cases the motivation – fame and fortune – is easy to discern.

But there is another group of pseudoscientists who contribute equally to the genesis of nonsensical ideas: scientists. Sometimes they are motivated by good old-fashioned greed and sometimes they are merely deluded. We must remember that science is no magic talisman, no golden path to enlightenment, it is practiced by people with the same foibles as the rest of us.

Consider the following excerpts from the writings of celebrated scientists.

> "This rod and the male and femail serpents joyned in the proportion 3:1:2 compose the three-headed Cerebus which keeps the gates of Hell. For being fermented and digested together they resolve and grow dayly more fluid ...and put on a green colour and in 40 days turn into a rotten black pouder. The green matter may be kept for ferment. Its spirit is the blood of the green Lion. The black pouder is our Pluto, the God of wealth."
>
> — Isaac Newton (describing alchemic experiments)

> "Among all the great men who have philosophized about this remarkable effect, I am more astonished at Kepler than at any other. Despite his open and acute mind, and though he has at his fingertips the motions attributed to the earth, he nevertheless lent his ear and his assent to the moon's dominion over the waters, to occult properties, and to such puerilities."
>
> — Galileo Galilei (in response to Kepler's belief that the moon can affect tides on Earth)

> "I have often heard it said that using the brain makes the eyes lighter in color."
>
> — Nikola Telsa (in response to a query about his grey eyes)

These quotations are stark reminders that all men, even the greatest of thinkers, are men of their times. They also remind us that the boundary between science and pseudoscience can be murky. Indeed, at one time alchemy and astrology were parts of mainstream science.

The quotations above are not cherry picked instances of great men making minor mistakes. Some of the best scientists of all time spent considerable portions of their lives on what we would now consider dubious pursuits. For example, the great German astronomer Johannes Kepler (1571–1630) earned his living as a court astrologer and constructed his third law of planetary motion in a laborious attempt to determine the "music of the spheres".

Isaac Newton (1642–1727) left manuscripts amounting to some 10 million words. About one half of these concerned religion,[4] 1 million words dealt with

---

[4]Newton believed in heretical anti-trinitarianism.

alchemy, and 3 million with mathematics and science. Presumably these numbers are an indication of the amount of time Newton spent on nonscientific and pseudoscientific topics.

Sir William Crookes (1832–1919) was a British chemist, inventor of the Crookes tube, and discoverer of the element thallium. Crookes was also a follower of the spiritualism that was popular in Victorian England, believing that he had witnessed levitation, phantoms, ghost writing, telekinesis, and other occult phenomena that "point to the agency of an outside intelligence".

Prosper-René Blondlot (1849–1930) was a respected French physicist known for his work on electromagnetism. Unfortunately he is remembered today for his discovery of "N-rays" in 1903, which were purported to emanate from most substances. Blondlot's experimental results were always at the edge of detectability and several notable physicists failed to reproduce them. Eventually Robert Wood exposed the self-deception by surreptitiously removing part of the experimental apparatus during a demonstration.

Nikola Tesla (1856–1943), famous inventor of the AC electric motor, became increasingly eccentric as he aged. While doing experiments in 1917 he observed unusual signals that he interpreted as radio communications from Mars. He spent many years working on a death ray and devoted the latter years of his life trying to invent a machine that could photograph thoughts on the retina of the eye (Fig. 3.1).

Figure 3.1: A newspaper representation of Tesla's thought camera.

The largely self-educated English physicist, Oliver Heaviside (1859–1925) made important discoveries in the field of electrical engineering and rewrote Maxwell's equation in their modern form (see Sect. 5.4). He became increasingly eccentric in his old age, reportedly using granite blocks as furniture and painting his nails pink. He stood nearly alone in denouncing Einstein's theory of relativity and many of his ideas were so eccentric that journals refused to publish them.

The great American chemist Linus Pauling (1901–1994), who won two Nobel Prizes and endured much abuse for daring to promote peace during the cold war, also succumbed to age-related eccentrism. In the late 1960s Pauling became convinced that vitamin C in high doses was a panacea, able to cure the common cold, atherosclerosis, and even cancer. His book *Vitamin C and the Common Cold* (1970) initiated the vitamin craze that continues to this day.

William Shockley (1910–1989), was an English-American physicist who invented the junction transistor and won the 1956 Nobel Prize. He founded Shockley Semiconductors to market transistors; this company formed the nucleus of what was to become Silicon Valley. Later in life Shockley caused a stir with speculation on race and eugenics that is widely regarded as unwarranted and racist.

In 1989 respected British chemist Martin Fleischmann (1927–2012) and American chemist Stanley Pons (1943–) stunned the scientific world by announcing that they had discovered a desktop process to produce limitless fusion energy (called "cold fusion"). Unfortunately, they had only succeeded in fooling themselves. The subsequent fallout saw the pair flee to France to continue their research at a private laboratory funded by Toyota. The laboratory ceased operations after 6 years. A small but vocal group continues to advocate for cold fusion to this day.

There is a real sense in which these examples are unfair: every scientist who has ever lived and who will live will think or say something stupid. These men just happened to do so in more spectacular fashion than usual. Some of the things they believed have long since been put down, while others linger on, distorting reality in their own ways.

3.2.2 Science and the Media

If ideas are to become dangerous it is not enough that people invent them, they must be placed in the minds of the public. Since the days of Gutenberg, this role has been played by pamphlets and newspapers. The past century saw great innovation in media and communication with the advent of movies, radio, and television. And of course, a defining characteristic of the current age is the ready access to information brought about by the construction of the internet and cell phone infrastructures. When this is coupled with the increasing industrialization of science, the flow of information, both right and wrong, makes Niagara look like a runnel (Fig. 3.2).[5]

A typical path from wild idea to pseudoscience starts with a deluded scientist; his research is reported in a technical journal; this is then reported in the media; which is then repeated on various internet sites; the subject is taken up by a B-list celebrity; cycles back to C-list media outlets; which various friends and acquaintances listen to; and finally passes to you and into your store of "common knowledge".

Every step on the information highway introduces distortion.

It is almost a theorem that each step of this process will introduce distortion of some sort. In particular, journalists and the media have different motivating factors

---

[5]There is an old joke: the rate at which books are being added to science shelves will soon exceed the speed of light, but this does not violate any principles of physics since no information is being transmitted.

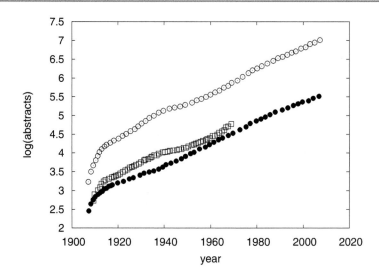

Figure 3.2: Cumulative science abstracts by year (chemistry: *circles*, physics: *squares*, mathematics: *filled circles*). The figure indicates that the rate of scientific publication is doubling about every 15 years.

Source: P.O. Larsen and M. von Ins, Scientometrics, **84**, 575–560 (2010).

from scientists: they seek exciting news for the purpose of selling newspapers. There is little room for caveats in this business.

> News item: "New research finds membership in a minority religion seems to hasten a loss of volume of the hippocampal region of the brain." Pacific Standard, May 10, 2011.

> The science: "This study examined prospective relationships between religious factors and hippocampal volume change using high-resolution MRI data of a sample of 268 older adults. …Significantly greater hippocampal atrophy was observed for participants reporting a life-changing religious experience. Significantly greater hippocampal atrophy was also observed from baseline to final assessment among born-again Protestants, Catholics, and those with no religious affiliation, compared with Protestants not identifying as born-again." A.D. Owen *et al.*, "Religious Factors and Hippocampal Atrophy in Late Life", PLoS one 6, e17006 (2011).

> Notice that the news outlet imputes causality, while the technical report does not.

> News item: "Cocaine floods the playground. Use of the addictive drug by students doubles in a year." Sunday Times, March 26, 2006.

The science: The study in question surveyed 9000 children from 300 schools and found the rate of cocaine use had gone from 1.4% in 2004 to 1.9% in 2005. These figures were rounded to 1% and 2% (hence the "doubling"). The figures were not highlighted by the researchers because they enquired after many dozens of drugs (see Sect. 2.4.4 for the look elsewhere effect).

News item: "Swimming too often in chlorinated water 'could increase risk of developing bladder cancer', claim scientists." Daily Mail Reporter, March 17, 2011

The science: "The most highly educated subjects were less exposed to chlorination by-products through ingestion but more exposed through dermal contact and inhalation in pools and showers/baths." G. Vastaño-Vinyals *et al.*, "Socioeconomic status and exposure to disinfection by-products in drinking water in Spain", Environmental Health, **10**, 18 (2011).

The study in question examines the correlation of socioeconomic status with exposure to disinfectants – it does not discuss cancer.

News item: "Why stilettos are the secret to shapely legs." The Telegraph, February 2, 2011.

The science: The actual study examined the relationship of anatomical heels with calves. Ahn An *et al.*, "Variability of neural activation during walking in humans: short heels and big calves", Biological Letters, **7**, 539–542 (2011).

News item: "Why silent types get the girl. Study finds that men who use shorter average word lengths and concise sentences are preferred, while men who use verbose language are deemed less attractive." The Telegraph, February 21, 2014

The science: "This study reports on male and female Californians' ratings of vocal attractiveness for 30 male and 30 female voices reading isolated words. ... These results suggest that judgments of vocal attractiveness are more complex than previously described." M. Babel *et al.*, "Towards a More Nuanced View of Vocal Attractiveness", PLoS one **9**, e88616 (2014).

Ex. Research shows that the brain's hemispheres tend to serve special functions: the left is logical and verbal, the right is spatial and intuitive. But it is easy to run too far with this dichotomy – almost all complex abilities make extensive use of both hemispheres. Consider then, the steady deterioration and distortion of information in the following sequence of events:

1. Psychologist Doreen Kimura of the University of Western Ontario in London Ontario, gives a conference talk reporting that melodies

fed to the left ear (hence the right brain) of university students were more readily recognized than those fed to the right ear.

2. The New York Times reports that "Doreen Kimura, a psychologist from London, Ontario, has found that musical ability is controlled by the right side of the brain".

3. A syndicated newspaper story says, "London psychologist, Dr. Doreen Kimura claims that musicians are right-brained."

4. A subsequent story reports, "An English psychologist has finally explained why there are so many great left-handed musicians."

Source: D.G. Meyers, *Psychology*, Worth Publishers, 2004.

Ex. A recent study on the accuracy of news coverage of climate science found that MSNBC overstated risks in 8% of climate-related news segments, 30% of CNN's segments were misleading, and 72% of Fox News Channel's segments misrepresented the science. Source: A. Huertas and R. Kriegsman, *Science of Spin?*, Union of Concerned Scientists report, April 2014.

While the majority of science reporting is reliable, these examples illustrate how things can go wrong. A common problem is overstating the case made by the scientists, who tend to be too cautious for the news. Our examples have focussed on rarer issues with misunderstanding original research.

Check your sources.

3.3 Case Studies

In 2003 it was discovered that Steve Jobs (1955–2011) had a rare and aggressive cancerous tumor in his pancreas. Doctors urged him to have an operation to remove the tumor, but Mr. Jobs demurred and instead tried a variety of new age and alternative remedies: a vegan diet, juices, herbs, acupuncture, he even consulted a psychic. Nine months later the tumor had grown and only then did he agree to surgery, which revealed that the cancer had spread to his liver. Although a liver transplant and other conventional medicine extended his life, he eventually died in 2011 due to a relapse of the pancreatic cancer.

Mr Jobs' decisions reflect a common distrust in mainstream medical science. Is this distrust deserved? Where does one draw the line between trust and mistrust? The many examples we have seen of pseudoscience show that this is a difficult question. In this section we seek to gain a better understanding of the phenomenon by making a more extensive study of some of the pervasive and long-lived examples of medical quackery.

It should not be surprising that pseudoscience tends to infest medicine. As we have seen, performing experiments on complex systems like the human body is very difficult. When this is coupled with large financial pressures and the personal interest we have in the well-being of loved ones, pseudoscience is bound to emerge (there is not much money, and even less love, in laser physics).

3.3.1 Pseudomedicine

Sometimes pseudomedicine (to coin a more specific term) is easy to spot, especially when it has fallen out of fashion. Consider the rather old-timey notions of eugenics or of *phrenology*. In case you have not heard of it, phrenology is the study of the shape and size of the skull in an attempt to learn something of human character and health. It was developed in 1796 by German physician Franz Gall (1758–1828) and was very popular in the 1800s (Fig. 3.3).

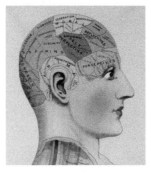

Figure 3.3: Phrenological model of the brain.

Many other fields of pseudomedicine have similarly long pedigrees. Among these are *chiropracty*, which was founded in 1895 by the Canadian-American physician Daniel Palmer (1845–1913). It is based on the belief that misalignment of the spine is the cause of all disease. Since then the field has (sometimes) repressed its unconventional origins, and chiropracty now largely deals with physical therapy.[6]

In a similar vein, osteopaths believe that manipulating muscles and joints help the body to heal. *Osteopathy* was created by American physician Andrew Still (1828–1917) in 1874 and is practiced in many countries today. Like chiropracty, the field has evolved and has lost some of its nonscientific tenets. Currently, the US medical community recognizes *ostepathic physicians*, who are medically trained, and *osteopaths*, who are limited to non-invasive manual therapies.

---

[6]Although a conversation with a physical therapist can reveal alarming remnants of the old modes of thought.

The Asian practice of ***acupuncture*** was essentially unknown in the west until the 1970s, when its popularity exploded (along with other exotic things like Hinduism, feng shui, qi, Ayurvedic medicine, and bubble tea). Acupuncture, like chiropracty and osteopathy, has no proven physiological benefits beyond the level of placebo. Nevertheless, it has been enthusiastically embraced by a population seeking alternative forms of medical treatment. The idea is that the stimulation of specific points on the body, usually by inserting a needle, corrects imbalances in the flow of ***qi*** (a natural energy not unlike animal electricity) through channels called ***meridians***. It hardly needs said that medical science has progressed significantly since this idea was invented, and no evidence exists for any of these concepts (Fig. 3.4).

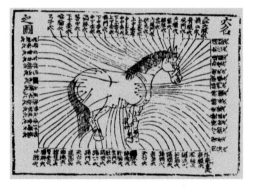

Figure 3.4: Acupuncture for farm animals.

3.3.2 Vitamins

Linus Pauling has been mentioned as the man who introduced the notion that massive doses of vitamins can cure a variety of ills (Sect. 3.2.1). Pauling and the vitamin industry succeeded brilliantly; as of 2012 more than one half of all Americans took some form of vitamin supplement, to the tune of $23 billion per year.

But what evidence is there? A recent article in *The Atlantic* reviewed studies on vitamin megadose regimens.[7] The author lists dozens of studies showing that large doses of vitamins can be detrimental to health. The litany ends with the startling statement, "In 2008, a review of all existing studies involving more than 230,000 people who did or did not receive supplemental antioxidants found that

---

[7]P. Offit, "The Vitamin Myth: Why We Think We Need Supplements", The Atlantic, July 19, 2013.

vitamins increased the risk of cancer and heart disease." This is serious stuff, a significant fraction of the population could be harming themselves with vitamins.

But, as we have stressed for two chapters, circumspection is required. Although the author of the article listed dozens of studies supporting his claim, it is possible that he was cherry-picking reports (selection bias and the look elsewhere effect!). Certainly, since vitamins occur naturally in food, at some sufficiently small dosage a vitamin regime must have no discernible effects on health. A quick consultation with the Cochrane Summaries revealed the following:

Vitamin C for preventing and treating the common cold

"This review is restricted to placebo-controlled trials testing 0.2 g/day or more of vitamin C. Regular ingestion of vitamin C had no effect on common cold incidence in the ordinary population, based on 29 trial comparisons involving 11,306 participants. However, regular supplementation had a modest but consistent effect in reducing the duration of common cold symptoms, which is based on 31 study comparisons with 9745 common cold episodes. Trials of high doses of vitamin C administered therapeutically, starting after the onset of symptoms, showed no consistent effect on the duration or severity of common cold symptoms."

Thus it appears that regular doses of vitamin C might reduce the duration of colds,[8] although it does nothing for catching colds. Unfortunately, there is no Cochrane summary for the treatment of cancer with vitamins, but I did find this:

Selenium for preventing cancer

"Recent trials that were judged to be well conducted and reliable have found no effects of selenium on reducing the overall risk of cancer or on reducing the risk of particular cancers, including prostate cancer. In contrast, some trials suggest that selenium may increase the risk of non-melanoma skin cancer, as well as of type 2 diabetes, raising concern about the safety of selenium supplements. Overall, no convincing evidence suggests that selenium supplements can prevent cancer."

In view of this and the studies mentioned above, the vitamin hype spread by a culpable and credulous media is not doing anyone any good:

"More and more scientists are starting to suspect that traditional medical views of vitamins and minerals have been too limited. Vitamins – often in doses much higher than those usually recommended – may protect against a host of ills ranging from birth defects and cataracts

---

[8]How were these durations determined? Is interviewer bias possible?

to heart disease and cancer. Even more provocative are glimmerings that vitamins can stave off the normal ravages of aging."

Anastasia Toufexis, Time Magazine, April 6, 1992.

At best this is entertainment; at worst it is dangerous and misleading tripe.

The practitioners of pseudomedicine are not content with the status quo. The budget for the National Center for Complementary and Alternative Medicine continues to increase, in part due to efforts by Senator Tom Harkin of Iowa. Senator Harkin is also introducing legislation to prohibit "discrimination" against practitioners.

Naturopaths have recently formed an association to accredit naturopathic medical colleges, and have convinced 18 states and the District of Columbia to grant licenses to naturopaths. And in an alarming turn, Washington joined Oregon and Vermont is covering naturopathic care under Medicaid.

> "This was a serious mistake. Naturopathic medicine isn't a coherent discipline like otolaryngology or gastroenterology. Take a look at the curriculum at Bastyr University, one of five U.S. schools accredited by the Association of Accredited Naturopathic Medical Colleges. Botanical medicine, homeopathy, acupuncture, and ayurvedic sciences are all on the course list. The only thing that these approaches to medicine share in common is that they lack a sound basis in evidence. Differentiating between trained quacks and untrained quacks is not actually a consumer protection measure."

Brian Palmer, "Quacking All the Way to the Bank", *Slate*, June 3, 2014.

3.3.3 Electromagnetic Fields and the Body

In 1780 Italian physician Luigi Galvani (1737–1798) discovered that applying an electrical charge to a frog's leg made it twitch. Galvani thought that the effect was due to an electrical fluid (which he called *animal electricity*) in the frog. This event initiated a long, continuing, and painful association of electromagnetic effects with animal physiology.

Galvani's contemporary and competitor, Alessandro Volta (1745–1827) thought that Galavani had not discovered something concerning animal electricity, but rather a property of salty water interacting with metal. He therefore set about constructing an artificial frog leg, eventually succeeding with his *electric pile* (battery). Although Volta's intuition proved correct, this did not diminish enthusiasm for the concept of animal electricity. Indeed, the imbuing qualities of life had always been a mystery; what better way to deal with it than to associate it with the new and equally mysterious electricity?

One of the most famous exemplars of the concept of animal electricity and a testament to its enduring appeal is Mary Shelley's (1797–1852) *Frankenstein*.

The novel is an allegory about the irresponsible use of power, but of course the driving element is the animation of Frankenstein's monster with electricity.

At the same time, a German physician named Franz Mesmer (1734–1815), developed the hypothesis that animate and inanimate matter could transfer energy, which he called *animal magnetism* (Fig. 3.5). It is likely that the theory was an offshoot of his thesis work, which postulated that the moon and planets created tides in human bodies that affect health. Unfortunately, Mesmer veered towards charlatanism and chose to apply his theory to curing susceptible young women of a variety of ills. The resulting scandal drove him from Vienna to Paris where he set up shop and continued his practice. Eventually King Louis XVI set up a commission chaired by Benjamin Franklin to investigate.

Figure 3.5: Mesmer affecting a cure.

Franklin's committee proceeded by impersonating Mesmer and watched as blindfolded patients expressed joy at feeling the effects of the treatment (which of course, only Mesmer or his helpers could administer). This may be the first time in medical history that a placebo was employed in an experiment. They then had actual healers "magnetize" subjects without the latters' knowledge, and noted that the treatment had no effect. In the *coup de grace*, the committee had a healer magnetize a tree and asked a 12 year old boy who was known to be sensitive to the procedure to locate the tree while he was blindfolded. After clasping several unmagnetized trees he went into convulsions and passed out. The committee's report of the spectacle is notable for its impeccable logic and for stressing the novelty and importance of the (to be named) placebo effect.

> "If the young man had felt nothing even under the magnetized tree,
> one could simply say that he was not very sensitive – at least on that

day. But the young man fell into a crisis under a nonmagnetized tree. Consequently, this is an effect which does not have an exterior, physical cause, but could only have been produced by the imagination. ... [This raises issues pertaining to] a science which is brand new, the science of the influence of the psychological on the physical."

P. Dray, *Stealing God's Thunder*, Random House, 2005.

The possible effects of electromagnetic radiation on the human body remain a rich source of speculation to this day. A quick perusal of a single medical text[9] on the subject reveals hundreds of studies with the formula "The Influence of Electromagnetic Fields on X", where X can be cellular and molecular biology, reproduction, the nervous system, thermoregulation, the cardiovascular system, hematology, immune response, orthopedics, galvanotaxis, cancer, or any other biological system or function. It is perhaps not surprising that many of these studies bear the stigmata of pseudoscience.

A popular current thread in the animal electricity saga is the idea that low energy electric and magnetic fields can adversely affect health. In the last 30 years the litany of suspects have included house wiring, microwave ovens, power lines, cell phones, wifi transmitters, smart electric meters, and airport scanners. We will discuss the physics behind these claims in detail in Chap. 5. In the meantime, we note that – perversely – people love electromagnetic fields! They are used in magnetic bracelets to "cure" a variety of ills, as scanners to evaporate body fat, and most ironically, the very same millimeter wave airport scanners that worry some people are used for therapy in Russia (in the quotation, MMW refers to millimeter wave):

> "MMW therapy is a widely used therapeutic technique that has been officially approved by the Russian Ministry of Health. In fact, it has been reported that, as of 1995, more than 3 million people have received this therapy at over 1000 MMW therapy centers in Russia. MMW therapy has been reported in the treatment of over 50 diseases and conditions." Source: K.L. Ryan *et al.* Health Physics, **78** 170 (2000).

The nonhuman branch of bioelectrics has dealt primarily with animal navigation. This idea has been current since at least 1855, when it was suggested that homing pigeons make use of the Earth's magnetic field as a navigational aid. The modern era can be dated from "Magnets Interfere with Pigeon Homing", by William Keeton (1933–1980). These days the idea that animals can detect magnetic fields for whatever purposes is called ***magnetoception***.

While it is possible that pigeons use the Earth's magnetic field as an aid to navigation (it must be supplemented with additional information since the knowledge

---

[9]See for example, *CRC Handbook of Biological Effects of Electromagnetic Fields* by C. Polk and W. Postow

of an absolute direction is not sufficient to navigate), it is becoming increasingly popular to claim magnetoception in animals for dubious purposes.

> Ex. "We demonstrate by means of simple, noninvasive methods (analysis of satellite images, field observations, and measuring "deer beds" in snow) that domestic cattle and grazing and resting red and roe deer, align their body axes in roughly a north–south direction. ... Magnetic alignment is the most parsimonious explanation."
>
> Source: S. Begali *et al.*, "Magnetic alignment in grazing and resting cattle and deer", PNAS, **105** 13451-13445 (2008).

One is forced to ask, for what purpose have cattle and deer evolved this ability? Surely there is no benefit to aligning in a particular direction while grazing – in fact one can easily come up with several evolutionary disadvantages.

> Ex. "We show that extremely low-frequency magnetic fields generated by high-voltage power lines disrupt alignment of the bodies of these animals with the geomagnetic field."
>
> Source: H. Burda *et al.*, "Extremely low-frequency electromagnetic fields disrupt magnetic alignment of ruminants", PNAS **106**, 5708–5713 (2009).

It is always worth one's while to check the size of the claimed exposure when evaluating claims such as this. The strength of a magnetic field 100 feet from a 230 kV power line is around 7 mG.[10] This should be compared to the Earth's magnetic field, which is around 400 mG. Is it plausible that a 1.8 % effect can completely disrupt the magnetic powers of a cow? Worse, the small magnetic field from the power line simply adds to the Earth's magnetic field, yielding a new slightly rotated field that the cow will still be able to detect. There should be no "disruption".

> Ex. "Dogs preferred to excrete with the body being aligned along the north–south axis under calm magnetic field conditions. This directional behavior was abolished under unstable magnetic fields."
>
> Source: V. Hart *et al.* "Dogs are sensitive to small variations of the Earth's magnetic field", Frontiers in Zoology, **10**, 80 (2013).

What possible evolutionary purpose can defecating in a given direction serve? If this increased survival fitness in dogs, why don't all mammals defecate while facing north? What do the authors mean by "unstable magnetic fields"? It is true that storms and the like cause fluctuations in the Earth's magnetic field, but these are tiny. How can a small exposure eliminate an outcome? (Answer: it can if the outcome is statistically insignificant or in error).

---

[10] Source: National Institute of Environmental Health Sciences.

Never underestimate how foolish pseudoscience can get.

3.3.4 Homeopathy

Homeopathy has a longer history than most of the "opathies" we have studied. Homeopathy is yet another alternative system for medicine that was created in 1796 by German physician Samuel Hahnemann (1755–1843) (Fig. 3.6). The field is based on his doctrine that "like cures like", which means that a substance that causes a particular disease in healthy people will cure sick people. While this may sound silly, one must remember that the germ theory of disease was 100 years away.[11]

Figure 3.6: Samuel Hahnemann (1755–1843), German physician.

Since he did not want to induce disease in his patients, Hahnemann proceeded by systematically diluting his medicines until positive outcomes were detected. He also "potentized" his medicines by "succussing" them, which means giving them a good shake.

Modern homeopathy has developed standards for dilution. For example a dilution of a substance by a factor of 100 is denoted 1C, while dilution by a factor of 100^2 is denoted 2C. These dilution factors typically reach ridiculous levels. A popular homeopathic treatment for the flu is a 200C dilution of duck liver, marketed under the name oscillococcinum. There are about 10^{80} atoms in the entire observable universe, a dilution of one molecule in the observable universe thus corresponds to 40C.[12] Oscillococcinum would thus require 10^{320} more universes

---

[11]Louis Pasteur made his ground breaking discoveries between 1860 and 1864.

[12]$40C = 100^{40} = 10^{80}$.

to contain a single molecule of active ingredient. These ludicrous numbers make it clear that homeopathic medicines contain no active ingredients and are therefore equivalent to placebos.

But one must have sympathy for Hahnemann – he was legitimately concerned with the well-being of his patients, and the atomic theory of matter was unknown. Furthermore, at the time mainstream medicine was primitive, relying on methods like bloodletting and purging, trepanation,[13] and poultices such as Venice treacle (Fig. 3.7).[14] Not surprisingly, these treatments often worsened symptoms and sometimes proved fatal. Hahnemann's rejection of these practices was a rational response to piecemeal and slipshod medical practice. Furthermore, by diluting his medications, Hahnemann replaced dangerous practices with benign placebo-based ones, thereby greatly improving the health outcomes of his patients! Unfortunately, the same cannot be said of modern practitioners of Hahnemann's ideas.

Figure 3.7: Detail from *The Extraction of the Stone of Madness*, by Hieronymus Bosch.

3.3.5 MMR Vaccine and Autism

In 1988 a British physician named Andrew Wakefield (1957–) and several of his colleagues published a study claiming that the MMR (measles, mumps, and rubella) vaccine causes autism and a terrifying new syndrome of bowel and brain damage in children. The study examined 12 children, of which 8 were supposed to have displayed symptoms an average of 6 days after having received the MMR vaccine. Because the article relied on parental reports of the subject's behavior, did not have a cohort, and did not have a control group, it would doubtlessly have

---

[13]Trepanation is the practice of drilling holes in the skull to treat disease.

[14]Venice treacle was made from 64 substances including opium, myrrh, and snake flesh.

sunk into the obscurity it deserved. However, Wakefield made statements at a press conference imputing causality in their observation and calling for suspension of MMR vaccination that fired the media into action.

The subsequent media storm led to a health crisis in the UK and soon spread to Europe and the US. The issue gained traction in the United States when it was taken up by actress Jenny McCarthy (1972–), whose book on the subject, *Louder than Words: A Mother's Journey in Healing Autism* was published in 2007. Further publicity was generated by an appearance on the Larry King Live show in 2008 and a PBS Frontline documentary in 2010.

Measles is no joke. Before the measles vaccination program started in 1963, 3 to 4 million people got measles each year in the United States. Of those people, 400 to 500 died, 48,000 were hospitalized, and 1000 developed chronic disability from measles encephalitis.

As a result of the media campaign, immunization rates in the UK dropped from 92 % to 73 % and it is estimated 10 extra children died. The effect was not nearly as dramatic in the United States, but researchers have estimated that as many as 125,000 American children born in the late 1990s did not get the MMR vaccine because of the controversy. As a result, measles cases in the US are increasing, despite the US being declared free of measles in 2000. Meanwhile, France was hit by a major outbreak of measles in 2011, which was blamed on the failure to vaccinate.

Wakefield's original study was examined in great detail by investigative reporter Brian Deer. It was eventually determined that the study was fraudulent: Deer found that Wakefield changed the data to such an extent that some of the children Wakefield claimed had autism were in fact perfectly healthy. Some of the children he said got sick after they were vaccinated were actually sick before the vaccinations. In no single case could the medical records be fully reconciled with the descriptions, diagnoses, or histories published in the study.[15] Finally, Wakefield was paid $750000 by lawyers trying to sue a vaccine manufacturer. In 2010 the study was withdrawn by the journal that published it and Wakefield was disbarred as a physician in Britain. He currently lives in Texas.

It is impossible for a single physician to generate an international health crisis – this could not have happened without the willful participation of a media that was determined to mine a scandal for sales (see Fig. 3.8). Part of the problem was that experienced science reporters were not involved in the story at many newspapers – because the scandal was "news", it was being handled by mainstream reporters. The saga is a visceral illustration of the importance of sober reporting and careful assessment of all extraordinary scientific claims.

---

[15]Source: B. Deer, "How the case against MMR vaccine was fixed", British Medical Journal, **342**, c5347 (2011).

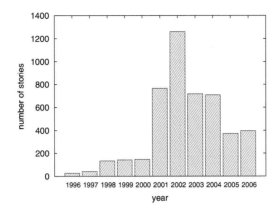

Figure 3.8: MMR news coverage in British newspapers.

Source: Ben Goldacre, *Bad Science*, Faber and Faber, 2010.

Extraordinary claims require extraordinary proof.

3.3.6   Intercessory Prayer

Invoking God's good will to help the sick has been a mainstay of religion for millennia. God's role as physician-in-the-sky is manifest at Lourdes, Fatima, and countless other sites that attract hordes seeking supernatural help for natural afflictions.

Can prayer help the ill? The question lies at the intersection of science and religion, of the ancient and the modern. At first sight, it appears to be scientific since it is testable and therefore falsifiable. But we will see that things are not so simple. Furthermore, the theology is even more suspect than the science. It turns out it is not easy living at an intersection.

The earliest attempt to answer the question was made by Sir Francis Galton (1822–1911) in 1872. Galton, a cousin of Darwin, was a Victorian gentleman with an epic range of interests and talent. One of a small cadre of polymaths that Europe and America seemed to throw up at will in the 19th century, Galton had a hand in founding or expanding psychology, geography, meteorology, genetics, statistics, forensics, and anthropology. Fascinated with his cousin's work on evolution, Galton flung himself into studies of human nature, and introduced the use of statistics and surveys in his research. Twin studies were first used by Galton in his research on inheritance, as were the phrases, *regression to the mean* and *nature versus nurture*. Galton also coined the word *eugenics* and was an early proponent of the idea of improving the lot of mankind through careful breeding.

Of course the idea and the field fell out of favor after it was appropriated by the Nazi intelligencia. These days the Galton Chair of Eugenics at University College London is called the Galton Chair of Genetics.

Galton's idea to test the efficacy of prayer was simple: royalty in the UK are regularly prayed for because they head the Anglican Church. He therefore undertook to compare the average lifetime of members of the royal family to other persons of wealth.

Galton prepared a table of average lifespans based on data he collected himself which showed no evidence for royal longevity, and hence, no evidence for the efficacy of prayer in promoting good health: "The sovereigns are literally the shortest lived of all who have the advantage of affluence. The prayer has therefore no efficacy, unless the very questionable hypothesis be raised, that the conditions of royal life may naturally be yet more fatal, and that their influence is partly, though incompletely, neutralized by the effects of public prayers."[16]

Although not definitive, one might think that Galton's study would have dampened any enthusiasm for further scientific tests. But it is estimated that five million dollars is spent annually on the subject, that about one half of American medical schools offer courses on spirituality and health, and that about 1000 articles are published per year on the subject.

We have already seen one example in this area (Sect. 3.2.2): recall that Dr Amy Owens of Duke University and colleagues decided to use MRI scans to check the correlation of brain structure with religiosity.

A more mainstream modern effort is provided by Dr W.S. Harris and colleagues who examined 990 patients who had been admitted to the Coronary Care Unit at the Mid America Heart Institute in Kansas City. The study was double blind and has a decent sample size and did indeed find a significant difference ($p = 0.04$) between outcomes of the group being prayed for over the control group.[17]

> "Remote, intercessory prayer was associated with lower CCU course scores. This result suggests that prayer may be an effective adjunct to standard medical care."

Although this seems definitive, the authors note

> "There were no statistically significant differences between groups for any individual component of the MAHI-CCU score. Mean lengths of stay in the CCU and in the hospital (after initiation of prayer) were

---

[16]Source: F. Galton, "Statistical Inquiries into the Efficacy of Prayer", The Fortnightly Review **12**, 125 (1872).

[17]Source: W.S. Harris *et al.*, "A Randomized, Controlled Trial of the Effects of Remote, Intercessory Prayer on Outcomes in Patients Admitted to the Coronary Care Unit", JAMA Internal Medicine **159**, 2273–2278 (1999).

not different, and median hospital stay was 4.0 days for both groups. There was no significant difference between groups using Byrd's hospital course score."

In other words, if you look at other ways to score the patients, no difference in outcomes is seen. This is, in fact, a serious issue in study design that we have seen before (hint: see Sect. 2.4.4). To prove the point, a 2006 study with 1802 CCU patients at six hospitals found no effect.

This is not a book on theology, but there seems to be something seriously amiss on the religious side of this issue. Surely it is sacrilegious to presume that we can explore the mind of God like he is a weighted spring in a middle school science experiment. If prayer studies made sense one could determine which religion God prefers, or how much the piety of the intercessor can influence God's actions, or how receptive God is to prayer as a function of distance.

But one person's tongue-in-cheek illustration is another person's science. In 2010, Dr Candy Brown and colleagues conducted a study with vision and hearing impaired subjects in rural Mozambique. Astonishingly, they found that *proximity* of intercessor and patient is correlated with health outcomes.[18] (For more on this see Ex. 10).

The press release for this study made a revealing statement:

> "Brown recounted that one subject, an elderly Mozambican woman named Maryam, initially reported that she could not see a person's hand, with two upraised fingers, from a distance of one foot. A healing practitioner put her hand on Maryam's eyes, hugged her and prayed for less than a minute; then the person held five fingers in front of Maryam, who was able to count them and even read the 20/125 line on a vision chart."

Unfortunately, this little anecdote reveals several violations of good study design – see if you can identify them.

If prayer studies are not good religion, they are not good science either. Omniscient and omnipresent deities need not respond in predictable ways to intercessory invocations. I am sure that the religious would say "God does not choose to reveal himself" in response to any outcome whatsoever from a prayer study. So what is really being studied?

3.4 Deconstructing Pseudoscience

It is possible that you believe some of the examples of pseudoscience that I have presented here. Maybe you take vitamin C when you have a cold and legitimately

---

[18]Source: C. Brown *et al.*, "Study of the therapeutic effects of proximal intercessory prayer (STEPP) on auditory and visual impairments in rural Mozambique", Southern Medical Journal, **103**, September 2010i.

feel better. Or maybe your aunt uses acupuncture to relieve back pain. That is fine, but remember that anecdote is not data and that the placebo effect is strong.

How is one to decide? Belief is built by obtaining information, evaluating its plausibility, and incorporating it into one's world-view and memory. The second step is the crucial one – people tend to get lazy and trust authority figures, or anything in print, or what their mother tells them. The way to improve things is to develop a common sense evidence-based understanding of the natural world that you can validate new information against. Furthermore, the third step in the process can have a surprisingly large effect on the second. Namely, one's worldview can blind one to new information. We examine both facets of belief-building in the following sections.

3.4.1   Detecting Pseudoscience

Developing a rational and healthy worldview requires some work on the part of the individual. Fortunately, there are many signs that point to pseudoscience in the making.

1. Violating Laws of Physics

Claims that violate laws of physics are an immediate sign of pseudoscience. It is common, for example, to see devices that extract energy from some source (water, the vacuum, hydrogen) that violate the law of conservation of energy. Often these devices come in the form of *perpetual motion machines* that supposedly can run forever. This, of course, violates the second law of thermodynamics.

2. Excessive and Baffling Jargon

Scientific jargon can be intense, but it is all well-defined and serves a useful purpose. If you see jargon that is not defined, or appears to exist solely to impress the reader, beware! Consider the following quotation from George Gillette's description of his "spiral universe" theory:

> "Each ultimore is simultaneously an integral part of zillions of otherplane units and only thus is its infinite allplane velocity and energy subdivided into zillions of finite planar quotas of velocity and energy."

> —— G.F. Gillette, as quoted in *Fads and Fallacies in the Name of Science*.

Or the following from a website promoting a new source of energy:

> "BlackLight has developed a commercially competitive, nonpolluting source of energy that forms a predicted, previously undiscovered, more stable form of hydrogen called "Hydrino".

> The SunCell plasma-producing cell invented to harness this funda-
> mentally new primary energy source as electrical output uses a cat-
> alyst to cause hydrogen atoms of water molecules to transition to
> the lower-energy Hydrino states by allowing their electrons to fall
> to smaller radii around the nucleus."

While this sounds pretty sciency, discussion of hydrinos should be enough to cause concern in the careful reader.

Hopefully, this website will leave no doubt as to its scientific plausibility:

> "From the center of your heart, emanates a field. It's the portal to true
> power: creating from the heart. This toroidal energetic field can reach
> as far as 5 miles from your heart. It's 5000 times more powerful than
> the brain, and affects everything in your life, there is a way to access
> it..."

3. Results Were "Published" Directly to the Media

Real scientists (mostly) develop an idea, test it, bounce it off of colleagues, test it again, write it up, send it out for review, publish it, and then, if it is exciting, issue a press release. As you can see, the idea has been heavily vetted by the time it reaches the public. Be wary if this process is bypassed.

4. Symptoms of Crackpottery

Some pseudoscientists are out to make money (see below) and do not care what people think of them or their ideas as long as cash is flowing in. On the other hand, a surprising number seek to have their inventions lauded by the scientific community.[19] Inevitably the approval does not arrive, his articles are not published in the mainstream journals, and prestigious prizes are not offered. How is a rational person to interpret the lack of recognition? It is simple: "they" are out to get him and seek to suppress his results because he is challenging their outdated theories. Beware of conspiracy, as pseudoscience lies nearby.

5. Someone Is Making Money

No surprise here riches are a powerful incentive and one should be wary when new theories are coupled with attempts to extract money. It has been argued that the largest scammers of this sort are the pharmaceutical companies, who (sometimes) make billions of dollars from heavily marketed, but essentially useless, drugs.

---

[19]The former pseudoscientist is called a "charlatan", the latter is a "crackpot".

6. Is There a Scientific Consensus?

Is the pseudoscientist an expert in the field in which he is making his claims? Is there a consensus about those claims? A good example is those who seek to deny that climate change exists. In general they are not climate scientists and, perversely, like to use data *generated by climate scientists* to "prove" their points. Of course, the odds that a climate scientist understands what his own data means is much greater than the odds a journalist or politician does.

7. Shoddy Studies

One can often access studies directly on the web. Often reading the abstract is enough to assign a Jadad score to the study. Check for biases, sample size, blinding, and randomization.

8. Problems with Claims

Pseudoscientists make great discoveries. However the effects they see tend to suffer from a number of issues: they are unreasonably accurate, error bars are not provided or measured, the effect is near the limit of detectability, the magnitude of the effect is independent of the intensity of the cause.

9. Criticism Is Met with Excuses

Pseudoscientists eventually run into opposition; how they behave in this circumstance is telling. Be aware if they exhibit symptoms of crackpottery (see above), become defensive, or make excuses for why their results cannot be reproduced.

10. Fantastic Theories Are Required to Explain the New Effect

The discovery is so unusual that it cannot be explained with mainstream science, and new scientific principles must be postulated. How do they know the new theories are true? The theories explain the data!

3.4.2   The Psychology of Belief

Old wive's tales, myths, vague impressions, rumor, fable, lore, legend – there is a lot of belief out there that is not supported by solid evidence. Why?

Certainly the quality of information that reaches the average person can sometimes be dubious. The business model of mass media must bear some responsibility for this. But this seems to exonerate the individual too much.

The cause may lie deeper: mainstream religion, spiritualism, and new age thinking all train their adherents to reject a rational interpretation of the universe at some level. Of course, if some things are accepted on faith, then it is an easy step to unquestioningly accept pseudoscientific claims. There is, in fact, a common idea that scientific progress is somehow damaging to the dignity of man.

Friedrich Nietzsche makes this point forcefully in *The Geneology of Morals* when he writes of "unbroken progress in the self-belittling of man" brought about by the scientific revolution. He goes on to mourn the loss "man's belief in his dignity, his uniqueness, his irreplaceability in the scheme of existence". It is no surprise that persons who subscribe to this attitude will tend to also subscribe to a wide array of nonscientific beliefs.[20]

This attitude underpins similar arguments that are heard often. For example, "science is arrogant" (and therefore should be discounted), or "science does not know everything, hence <insert favorite topic> is possible". Of course if the topic violates a law of physics, then this argument is not correct.

It is tempting to say that the problem is rooted in general scientific illiteracy. In some political circles it is fashionable to blame this illiteracy on public schools. Of course, sometimes our politicians *choose* to have poor schooling – in 1999 the Kansas State Board of Education decided to remove all mention of the Big Bang, radioactive dating, continental drift, the age of the Earth, and evolution from the state science education standards. More recently, Wyoming and Oklahoma have rejected the Next Generation Science Standards because they mention climate change. But is it true that Americans are scientifically illiterate? The US consistently ranks in the upper half of advanced countries in PISA testing on mathematics and science.[21] And American acceptance of science is very high, with 84 % agreeing that the effect of science on society has been mostly positive.[22] Once again, this reason appears to be too facile.

An important hint is provided by recent polls which show that Americans who do not believe in anthropogenic climate change are otherwise quite knowledgeable about science. This seemingly paradoxical situation is illustrated in the poll results of Table 3.2. The right two columns show the percent of left-leaning or right-leaning people who agree with various interpretations of climate change. A clear difference of opinion based on political stance is evident.

The interpretation of these results seems clear: one's values can trump evidence. Every person has a set of priorities, standards, and values that guides everyday decision making (which we referred to as a worldview in Sect. 3.4). New information that disagrees with these values tends to be discounted or outright dismissed, while information that reinforces one's worldview is readily accepted into the framework.[23] Personal values tend to be built on religion, politics, and science. And of these, science is the loser when it conflicts with the others.

---

[20]One could just as well argue that it is mankind's increasing understanding of nature that lends us our dignity.

[21]PISA is the Programme for International Student Assessment – a respected international testing program for 15-year-olds.

[22]Source: Pew Research poll, "Public Opinion on Religion and Science in the United States", Nov 5, 2009.

[23]See A. Corner *et al.*, "Public engagement with climate change: the role of human values", WIREs Climate Change, **5**, 411–422 (2014).

Table 3.2: US views on global warming.

Warming issue	Total (%)	Left (%)	Right (%)
Solid evidence			
The Earth is warming	67	84	46
(Due to human activity)	44	64	23
(Due to natural patterns)	18	17	19
(Don't know)	4	4	3
No evidence for warming	26	11	46
(Don't know enough)	12	7	20
(Not happening)	13	4	25
(Don't know)	1	0	1
Mixed evidence	7	5	7

Source: Pew Research Center Poll, Oct 2013.

REVIEW

Important concepts:

Nonscientific beliefs are common.

Ten signs of pseudoscience.

Pseudoscience is often started by scientists.

Extraordinary claims require extraordinary proof.

The media has different priorities and distortion is inevitable.

Most pseudoscience fades away, but some is very long lived.

Beliefs can be determined by nonscientific factors.

FURTHER READING

Martin Gardner, *Fads and Fallacies in the Name of Science*, Dover Publications, 1957.

Ben Goldacre, *Bad Science*, Faber and Faber, 2010.

Robert Park, *Voodoo Science: the road from foolishness to fraud*, Oxford University Press, 2000.

Carl Sagan, *The Demon-haunted world: science as a candle in the dark*, Ballantine Books, 1996.

EXERCISES

1. Devra Davis has written a book called *Disconnect: the truth about cell phone radiation, what the industry has done to hide it, and how to protect your family.* Do you see any signs of pseudoscience in the book title?

2. Look at Fig. 3.2, how many more chemistry abstracts are published than mathematics? Estimate the total number of chemistry abstracts.

3. CAM.

   Visit the web site of the National Center for Complementary and Alternative Medicine (nccam.nih.gov). Discuss what you find in view of what you have learned in this chapter.

4. AIDS deaths.

   Read the article *Mbeki aids denialism costs 300000 deaths* published in the Guardian in 2008 (tinyurl.com/o72tcgg). Comment on the article in light of what you have learned in this chapter.

5. Quantum Science.

   Visit the web site for "Quantum Jumping" (www.quantumjumping.com) and analyse what you see.

6. Dead Salmon.

   Read the article *Scanning Dead Salmon in fMRI Machine Highlights Risk of Red Herrings* from Wired magazine at tinyurl.com/ ls42c34. Comment on the report in light of what you have learned in this chapter.

7. Water Powered Cars.

   Read the article *Craving Energy and Glory, Pakistan Revels in Boast of Water-Run Car* in the New York Times (tinyurl.com/ nyvsa76). Comment on the article in light of what you have learned in this chapter.

8. Confused Birds.

   Read the article *Cracking Mystery Reveals How Electronics Affect Bird Migration* in National Geographic (`tinyurl.com/qy69g79`). Comment on the article in light of what you have learned in this chapter.

9. Homeopathic Drugs.

   The homeopathic drug "arnica montana" is rated 30C. By what factor has the active ingredient been diluted? If an arnica montana pill consists of 12 g of carbon (they are not very tasty), how many pills would you need to take to ingest one molecule of arnica montana? How much would they weigh in kilograms? How much would they weight in units of the Earth's mass?

10. Proximity in Prayer Studies.

    A commentary accompanied the proximity study mentioned in Sect. 3.3.6 by Dr John Peteet of Harvard. Dr Peteet was open-minded, writing[24]

    > "But the conviction that we live in a closed system governed only by naturalistic processes is an expression of faith in a world view rather than a conclusion logically demanded by the scientific method."

    (a) Does the author think prayer is a "naturalistic process"? How can prayer be tested if it is not?

    (b) What does the author mean by "closed system"?

    (c) What is he trying to say concerning "logical demands"?

---

[24]Source: J. Peteet, Southern Medical Journal, **103**, September 2010.

Energy and Entropy

"Energie is the operation, efflux or activity of any being: as the light of the Sunne is the energie of the Sunne, and every phantasm of the soul is the energie of the soul."

— Henry More, *Platonica (1642).*

The concepts of energy and entropy are crucial in the development of a quantitative and predictive science. They also provide the basis for much of the discussion to follow. The most important aspects of the economy and the environment can be traced to the flow of energy through these systems. Entropy, on the other hand, provides an important constraint on how this energy can flow. We will apply these concepts to the human body and the economy in this chapter, to the environment in Chap. 7, and to space travel in Chap. 10.

4.1 Definitions and Units

Like mass, it is not possible to define energy in terms of other quantities and operational definitions must be employed. Intuitively, *energy represents the ability to do things*. All of these "things" boil down to imparting energy to other objects. Thus a fire can heat water, a thrown brick can break a window, a moving plunger can depress a spring, and so on.

4.1.1 Kinetic and Potential Energy

Energy comes in two forms

kinetic energy: The ability to do things due to motion.

potential energy: The ability to do things due to position.

A slowly moving brick does less damage than a quickly moving one. Extensive experimentation and Newton's laws of motion indicate that a brick that is moving twice as fast can do four times the damage, while one that is twice as heavy does twice the damage. These relationships are summarized in the equation for (nonrelativistic) kinetic energy

$$KE = \frac{1}{2}mv^2, \qquad (4.1)$$

which you might recall from high school physics.

Potential energy represents the ability to do something due to the position of a body. Position with respect to what? The answer is something called a *field* that we will learn about in Chap. 5. For now just think of it as something related to the force acting on the body.[1] The ability to do something due to a body's position is only apparent once the body starts to move. When this happens we speak of potential energy converting to kinetic energy.

The concepts of potential and kinetic energy are somewhat fluid because one form of energy can change to another depending on how closely one is examining a system. We say that the forms of energy are *scale dependent*. For example, chemical potential energy is a large-scale concept, and closer examination reveals that it is a sum of atomic-scale potential and kinetic energy.

4.1.2 Conservation of Energy

Recall from Chap. 1 that the sum of kinetic and potential energy is conserved. In this sense the potential energy at a point is *defined* as the negative of the kinetic energy of a body at that point.

> Ex. A brick is held 50 feet above ground level where its potential energy is defined to be zero. Its kinetic energy is 100 J after falling 3 feet. Thus the potential energy 47 feet above ground level is defined to be -100 J.

Because total energy is conserved, one cannot speak of the creation or destruction of energy – only its transformation from one form to another. Consider for example, a pitcher throwing a baseball to his catcher. A baseball moving at 100 mph clearly has a lot of energy (see Ex. 3). This is kinetic energy that was imparted to the ball by the arm of the pitcher, which also has a lot of kinetic energy. The kinetic energy of the pitcher's arm comes from food energy, which is potential and kinetic energy associated with chemicals in the body. This biochemical energy was created in plants by converting the energy in sunlight, and that comes from nuclear reactions in the sun (see Chap. 8). Ultimately the energy that drives the sun comes from the creation of the universe, which – you will not be surprised to read – we are somewhat fuzzy about. The ball stops moving once it is caught, since energy does not disappear, the baseball's kinetic energy must go

---

[1]"Position" can be quite general. For example, an electron has a property called *spin* that is analogous to the spinning of a ball. The direction of spin is part of the electron's "position".

somewhere. I can think of four sinks for this energy: (i) air friction (i.e., the ball heats up the air, which means it imparts kinetic energy to air molecules) (ii) friction in the catcher's glove due to the impact of the ball (iii) recoil in the catcher's arm is converted to heat (iv) the sound of the ball hitting the glove takes up some of the energy of the ball.

This long chain of events started with the creation of the universe (!) and ended with energy being dissipated as heat or sound (which also eventually is dissipated as heat). Such is the fate of all the energy in the universe.

> Ex. A running refrigerator stands in an insulated room with its door open. Does the air temperature in the room go up, stay the same, or go down?
>
> A: The refrigerator is importing electrical energy via its plug, this energy must go somewhere, and ends up heating the room.

4.1.3 Units of Energy

A number of operational definitions of energy have been developed historically.[2] The SI unit is called the joule (symbol "J") in honor of James Prescott Joule (1818–1889), an English physicist and brewer who found the relationship between the mechanical expenditure of energy and heat. It was this work that led Hermann von Helmholtz to declare that energy is conserved in 1847. Joule's conclusions were formed after many years of painstaking experimentation. At issue was the fate of energy when it seemed to disappear (i.e., was dissipated into heat). This, of course, raises the question of the nature of heat, which was thought for centuries to be a type of invisible fluid called *caloric*. Joule's experiments (see Fig. 4.1) showed that it was possible to convert mechanical energy into heat in a measurable fashion and led him to speculate that heat was in fact a form of molecular motion (he imagined it to be rotational, which is sensible since otherwise one would expect objects to fly apart).

Since energy is defined operationally, all of the units of energy are given in terms of the energy required to perform a certain task. The most common ones are listed here.

joule (J) The energy required to accelerate a 1 kg mass for one meter at 1 m/s^2. Thus $1 \text{ J} = 1 \text{ kg m}^2/\text{s}^2$. It takes about 1 J to lift a billiard ball one meter.

erg An erg is like a joule except that the meter is replaced by a centimeter and the kilogram is replaced by a gram. Since $1 \text{ kg} = 1000 \text{ g}$ and $1 \text{ m} = 100 \text{ cm}$, $1 \text{ erg} = 10^{-7}$ joule.

small calorie (cal) The energy needed to raise one gram of water one degree Celsius.

---

[2] "For those who want some proof that physicists are human, the proof is in the idiocy of all the different units which they use for measuring energy." – Richard Feynman.

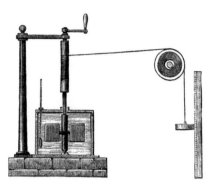

Figure 4.1: Joule's device for measuring the mechanical equivalent of heat. The falling mass to the right spun paddles in a water filled flask, and the resulting increase in temperature was measured with a very accurate thermometer.

large calorie (Cal) 1000 calories. This is the unit commonly used to describe food energy.

electron volt (eV) The energy gained when an electron moves through an electric potential difference of one volt.

British thermal unit (btu) The energy needed to heat one pound of water one degree Fahrenheit.

Since all of these units describe the same fundamental thing, they can be converted into each other by multiplying by the appropriate conversion factors. These are:

$$1 \text{ ev} = 1.6021 \cdot 10^{-19} \text{ J}, \tag{4.2}$$

$$1 \text{ Cal} = 4184 \text{ J}, \tag{4.3}$$

and

$$1 \text{ btu} = 1055.87 \text{ J}. \tag{4.4}$$

Ex. Convert 300 btu into electron volts.

$$
\begin{aligned}
300 \text{ btu} &= 300 \text{ btu} \times 1055.87 \frac{\text{J}}{\text{btu}} \times 6.242 \cdot 10^{18} \frac{\text{eV}}{\text{J}} \\
&= 1.98 \cdot 10^{24} \text{ eV}.
\end{aligned}
$$

Thus the energy required to raise 300 pounds of water one degree Fahrenheit is equivalent to the energy required to accelerate $2 \cdot 10^{24}$ electrons through an electric potential difference of one volt.

In future chapters we will sometimes refer to masses (typically of fundamental particles) in energy units. Although this may seem odd, the conversion can be made with the aid of Einstein's famous formula $E = mc^2$. Recall that a joule is $1\,\mathrm{kg\,m^2/s^2}$, so dividing an energy by a speed squared does indeed give a mass. A typical conversion factor is

$$1\,\mathrm{GeV}/c^2 = 1.783 \cdot 10^{-27}\,\mathrm{kg}. \tag{4.5}$$

This can be obtained as follows:

$$
\begin{aligned}
1\frac{\mathrm{GeV}}{c^2} &= 1 \cdot 10^9 \frac{\mathrm{eV}}{(3.0 \cdot 10^8\,\mathrm{m/s})^2} \times 1.6021 \cdot 10^{-19}\frac{\mathrm{J}}{\mathrm{eV}} \times 1\frac{\mathrm{kg \cdot m^2/s^2}}{\mathrm{J}} \\
&= 1.783 \cdot 10^{-27}\,\mathrm{kg}.
\end{aligned}
$$

4.1.4 Power

How quickly energy is transformed is an important concept – fast transformations are often called explosions! The rate at which energy is transformed from one form to another is called **power**. As with energy, many ways for defining power have been used.

watt (W) The Watt, named after Scottish engineer James Watt (1736–1819) is the SI unit of power. One watt is a joule per second (W = J/s).

horsepower (hp) The horsepower was introduced by James Watt to compare the power output of his steam engines to draft horses. Watt defined one horsepower as the amount of power required from a horse to pull 150 pounds out of a hole that was 220 feet deep in one minute.

ton The energy required to melt 2000 pounds of ice in 24 h.

Although the horsepower and ton may appear somewhat eccentric, these are all equally valid representations of the rate of change of energy. Of course one can use any of the units of energy defined above to make other definitions. For example electron volts per hour, ergs per month, or btus per year are all valid.

Conversely, one can multiply a unit of power by a time to obtain a unit of energy. Although this seems counterproductive, the kilowatt-hour is in common usage as a unit of energy.

Ex. 1 kWh = 1000 W × 1 hr = 1000 J/s × 60 × 60 s = 3.6 MJ.

Again, the units of power can be converted into one another by multiplying by appropriate factors. For the three we have focussed on one has

$$1\,\mathrm{ton} = 3504\,\mathrm{W}, \tag{4.6}$$

and

$$1 \text{ hp} = 746 \text{ W}. \tag{4.7}$$

Ex. A person consumes about 2000 Calories per day. Using the conversion factors we determine that this is equivalent to

$$2000 \frac{\text{Cal}}{\text{day}} \times 4184 \frac{\text{J}}{\text{Cal}} \times \frac{1}{24 \times 60 \times 60} \frac{\text{day}}{\text{s}} = 96.7 \text{ W}.$$

A person is a 97 W light bulb.

Be careful to distinguish the technical meanings of energy and power from those in common usage! For example a "power plant" is really an "energy plant".

4.2   Entropy

If you open a bottle of perfume in the center of a large room and come back a day later you will find that the bottle is empty and that the room smells of perfume. What has happened? At a microscopic level perfume molecules bounce around at a rate that is determined by the temperature in the room. On occasion one of these molecules will escape the bottle and continue its life bouncing among the air molecules. The odds of it bouncing back *into* the bottle are exceedingly small. This process continues until all of the perfume molecules are uniformly distributed about the room. In principle, if one waited long enough it is possible that all of the perfume molecules will bounce back into the bottle, but this would require waiting far longer than the age of the universe.

A useful way to think of this is as an increase of the *disorder* of the air-perfume system. When the experiment started, we could say with certainty that all molecules of perfume were in the bottle. In this sense, the system is *ordered*. At the end of the experiment one can only say that the perfume molecules are in the room, thus the order has decreased and the disorder has increased.

Increasing disorder is central to much of common experience – breaking a glass, burning wood, consuming food, turning on a light, slowing a bicycle. Because of its importance, the idea was formalized as **entropy** by Rudolph Clausius (1822–1888) in 1865. Clausius summarized the *second law of thermodynamics* (see Chap. 1 for more discussion) with the statement

---

the entropy of the universe tends to a maximum.

For our purposes we can restate this as "disorder increases" or "heat is created". The connection to the production of heat is a common one because disorder that is created is often in the form of random molecular motion associated with heat. For example, braking your car is a process that takes very organized motion (the car moving forward) and converts it into the disorganized motion of heat (in your brake pads and rotors).[3]

As disorder in an isolated system increases, the increase in entropy slows down and stops. We say that the system has reached *equilibrium*. Roughly, equilibrium means that nothing is happening, or can happen. A gas in equilibrium, for example, has about the same number of molecules with the same average speed in each subvolume. A more practical definition of equilibrium is that no energy can be expended, hence nothing useful can be done.

What happens to energy as entropy increases? The total energy (in an isolated system) stays constant, but in general potential energy is converted into kinetic energy. Since potential energy permits things to happen, it is often said that the *quality* of the energy has been degraded when it makes the transition from potential to kinetic. It is the business of power companies to make high quality energy (namely electricity) that is eventually converted to low quality energy in the form of waste heat. Said another way, high quality energy has low entropy, while low quality energy has high entropy.[4]

The second law of thermodynamics means that it is impossible for any process to continue indefinitely. For this reason it is not possible to build a "perpetual motion machine". This does not stop people from trying of course – attempts have been recorded as far back as 1159. Some 300 years later Leonardo da Vinci was irritated enough to proclaim, "Oh ye seekers after perpetual motion, how many vain chimeras have you pursued? Go and take your place with the alchemists." Many of the proposed designs are clever, but inefficiencies in any device mean that its energy must eventually dissipate. Charlatans continue to promote perpetual motion machine today. One recent example is the subject of Ex. 7.

4.3 Energy, Entropy, and Life

The body is a complex machine that extracts energy from the environment to perform a bewildering array of biochemical processes. Because these processes are highly ordered, all life reduces entropy locally. The second law of thermodynamics then implies that living things increase the entropy of the environment. This behavior, and therefore life, is impossible in systems in equilibrium. But, of course, the Earth is far from an equilibrium system since the sun supplies it with vast amounts of energy.

---

[3]Entropy can be related to *information* in a quantitative way. The connection is provided by the partial knowledge that we typically have of a system. For example, one may know the volume, temperature, and pressure of a gas, while having no idea of the positions and speeds of the roughly 10^{23} particles that make it up. In this view entropy is the amount of additional information needed to specify the physical state of a system.

[4]See Sect. 8.9.2 for an application of entropy to nuclear decay.

Table 4.1: Basal metabolic rates.

Organ	Power (W)	O_2 consumption (mL/min)
Heart	6	17
Kidney	9	26
Skeletal muscle	15	45
Brain	16	47
Other	16	48
Liver and spleen	23	67
Total:	85	250

The energy that drives cell growth, DNA replication, amino acid synthesis, enzyme kinematics, muscle use, and all the rest is supplied by potential energy contained in the food we eat. The rate at which energy is expended by various parts of the body is shown in Table 4.1. The figures are for an average person at rest (termed the **basal metabolic rate**). About 75 % of daily calorie use goes to these "internal" needs. The rest is expended on mechanical work, such as walking.

The table also shows the rate at which oxygen is consumed by each organ. Power and oxygen consumption track each other, indicating that it is oxidation that produces energy in the body. Thus it is possible to measure the energy content of food simply by burning it, which gives the familiar caloric values of food.

As you are aware, if more calories are ingested than are burned the excess is converted into fat and stored. The energy content of common food components is

Fat	9 Cal/g
Protein	4 Cal/g
Carbohydrates	4 Cal/g
Sugars	4 Cal/g.

For comparison Table 4.2 shows typical power expenditures for different activities.

Ex. You have eaten a 200 calorie apple. This corresponds to a weight gain of 200/9 = 22 g = 0.78 oz.
Q: What assumption has been made to obtain this figure?

Ex. To burn off the 200 calorie apple you just ate you would need to cycle for

$$200 \text{ Cal} \times 4184 \frac{\text{J}}{\text{Cal}} \times \frac{1}{400} \frac{\text{s}}{\text{J}} \times \frac{1}{60} \frac{\text{min}}{\text{s}} = 35 \text{ min.}$$

Table 4.2: Energy and oxygen usage rates by activity.

Activity	Energy (W)	O_2 (mL/min)
Sleeping	86	250
Sitting at rest	120	340
Thinking	210	600
Walking (5 km/h)	280	800
Cycling (15 km/h)	400	1140
Shivering	425	1210
Playing tennis	440	1260
Climbing stairs	685	1960
Cycling (21 km/h)	700	2000
Running cross-country	740	2120
Cycling (professional)	1855	5300
Sprinting	2415	6900

It might surprise you that you have to cycle for so long just to burn off an apple. Most foods are high in calories and can run a human body for a long time. If you are trying to lose weight it is far easier to not eat the apple in the first place.

Occasionally you will see a news item on a dangerously obese child. Invariably the distraught parents don't know how the child became obese or what to do about it. In terms of energy flow the answers are quite clear. To gain 100 pounds in 2 years you need to consume

$$\frac{100 \text{ pounds}}{2 \text{ years}} \times \frac{1000}{2.2} \frac{\text{g}}{\text{pounds}} \times 9 \frac{\text{Cal}}{\text{g}} \times \frac{1}{365} \frac{\text{years}}{\text{day}} = 560 \text{ Cal/day}$$

more than your body burns. This is four cans of pop, or two Snickers bars, or four ounces of potato chips per day. Viewed in this stark light, weight gain is not magical – obese children get that way because their parents facilitate it.

With world wide obesity rates climbing there is a tendency to look for quick fixes such as diet pills. There are only two ways that intervention such as this can work: the pills can make it harder to digest food or they can raise your metabolic rate. The first option can disrupt normal digestion, and it seems far safer to simply eat less. Chemically resetting a metabolic rate seems either hopeless or dangerous. Ingested energy must go somewhere – either one exercises more, or one becomes warmer. Presumably diet pills are not required if one is going to exercise more, so the only way they can contribute to weight loss is by increasing one's internal temperature. Doing this (if it is even possible) would surely require some serious manipulation of the body's hormonal system and, if it is to lead to noticeable effects, must therefore be dangerous. Of course, the simplest possibility is that diet pills do nothing.

4.4   Energy Flow in Society

4.4.1   US and Global Energy Consumption

Energy is required to extract the materials that create wealth. In fact every sector of the economy is dependent on energy, sometimes lots of it. Figure 4.2 shows the historical energy consumption of the USA. The units are "quads", which are 10^{15} btus, which is 10^{18} J, or 1 EJ. The country ran on wood up to about 1860, after which coal dominated for 100 years. Currently the nation's energy use of around 100 EJ/yr is provided by petroleum (36 %), natural gas (27 %), coal (18 %), and uranium (8 %). Importantly, 81 % of the nation's energy is provided by burning carbon. The implications of this will be discussed in Chap. 7.

Figure 4.3 shows where all of this energy goes, broken down by the broad areas of transportation (28 %), industry (25 %), commerce (8.8 %), and residential use (11 %). The massive amount of energy consumed in moving goods and people indicates how cheap transportation (i.e., petroleum) is and how much we value exotic goods. It is only within the last generation that global transportation has become so cheap that is possible to buy fruit from Brazil, electronics from China, cars from Japan, and clothes from Bangladesh as if it were commonplace.

Figure 4.2: Historical energy consumption in the USA.
Source: US Energy Information Administration.

Figure 4.3 also reveals that a shocking 61 % of imported or generated energy is wasted. The main culprits are electricity production and the transportation system that waste a whopping 67 % and 79 % of the energy they consume respectively. Viewed on this scale, attempts at energy conservation based on home insulation or more efficient light bulbs are laughable. We will discuss what can be done about this in Chap. 9.

When averaged over the entire population, 95 quads per year works out to about 10 kW per person. This is an enormous personal power. A healthy well-fed laborer can output about 75 W in an eight hour work day, so this corresponds to

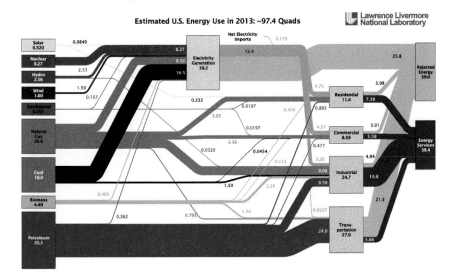

Figure 4.3: Energy flow in the US economy.

Source: Lawrence Livermore National Laboratory.

having 133 personal servants. Think what this means for your personal wealth. One hundred years ago my great grandparents *were* personal servants, now I control the equivalent of 133 of them.

Cheap energy makes us rich.

Global energy consumption in 2008 was 15 TW, which corresponds to 2 kW per person. It will come as no surprise that this power is distributed very unequally (Fig. 4.4). Can such disparity be sustainable? How shall we deal with increasing energy demands in the developing world as energy becomes scarcer? These are serious societal issues that must be faced at some time.

4.4.2 The Green Revolution

Plants are the primary consumer of energy on Earth. It is estimated that the total energy captured by photosynthesis in green plants is $2 \cdot 10^{23}$ joules per year. This corresponds to a power of 6 PW, which is about 400 times what humans produce.

This energy is mostly used to extract carbon from the air to build plant bulk. Figure 4.5 shows this activity in terms of the gross production of carbon per square meter per year. An eyeball average of $1 \, kg/m^2$ carbon consumption over the land surface of the Earth yields a consumption rate of 10^{14} kg per year. This is part of the *carbon cycle* that will be discussed in Chap. 7.

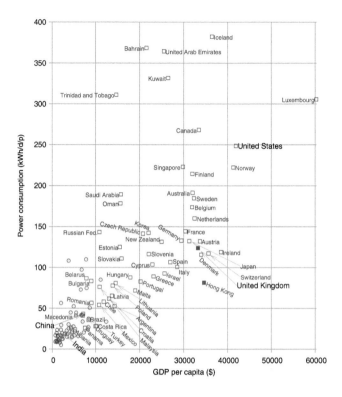

Figure 4.4: Per capita energy consumption vs. per capita GDP.

Source: D. MacKay, *Sustainable Energy – without the Hot Air*, published by UIT:
www.uit.co.uk/sustainable.   Also available for personal non-commercial use at
www.withouthotair.com.

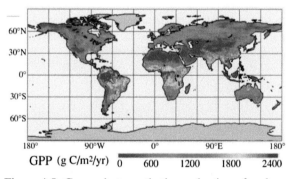

Figure 4.5: Gross photosynthetic production of carbon.

Source: A. Ito and T. Oikawa, *Global Environmental Change in the Ocean and on Land*, Eds.
M. Shiyomi *et al.*, 343–358 (2004). Reproduced with permission.

For much of the world it is the energy content of food that is of interest. This is, of course, because the food supply in many places is far from secure. Nevertheless, it is much better than it was – 70 years ago mass famines were depressingly common. Several advances came together in the 1950s to change this:

- biologists bred high-yield disease-resistant grains

- effective pesticides came into common use

- nitrogen fertilization came into common use.

The net result has been named the "Green Revolution". Its chief proponent was American biologist, Norman Ernest Borlaug (1914–2009), who has been called "The Man Who Saved A Billion Lives", with some justification (Fig. 4.6). During the 1950s Borlaug introduced disease-resistant wheat to Mexico. By 1964 the Mexican wheat harvest had sextupled, and Mexico became a net exporter of wheat. Because of his rising fame, Borlaug was invited to India in 1962 to assist in dealing with persistent threats of famine. His efforts led to a doubling of wheat yields in India and Pakistan between 1965 and 1970, and Pakistan became self-sufficient in wheat production in 1968. By 2000, wheat production had gone up by more than a factor of three in developing countries. Borlaug was awarded the 1970 Nobel Peace Prize for his contributions to the well-being of mankind.

Figure 4.6: Norman Borlaug.
Reproduced with permission from CIMMYT.

Much of the progress wrought by the Green Revolution was due to the addition of energy into agriculture via fertilizers. And of course, this energy derives from fossil fuels. It is estimated that 3 % of global energy use goes into agriculture. In the USA, 14 % of total economic energy use is taken up by the food chain (this includes farming, fertilizers, pesticides, transportation, refrigeration, and sales).

Simply growing food in North America is an energy-intensive business. Table 4.3 shows the energy content of various typical crops and the energy required to produce them. Grains need about ¼ of the energy they actually produce.

Table 4.3: Energy content in various crops.

Crop	Energy content (GJ/ha)	Added energy (GJ/ha)	Ratio
Rye	37.4	9.9	0.26
Oats	42.9	10.6	0.25
Canola	43.5	11.6	0.27
Soya beans	44.9	6.6	0.15
Barley	54.6	12.3	0.22
Winter wheat	78.8	13.8	0.18
Tame hay	82.2	4.6	0.06
Grain corn	116.7	18.4	0.16
Switchgrass	171	7.9	0.05

Source: J. Woods *et al.*, "Energy and the Food System", Phil. Trans. R. Soc. B, 365 (2010).

The situation is even worse for meat, as illustrated in Table 4.4. In this case as much as three times as much energy is required as is contained in the final product. Unless things change drastically, energy will become more expensive in future and consuming meat will represent a serious drain on the food supply.

Table 4.4: Energy content in various farm products.

Product	Energy content (GJ/t)	Added energy (GJ/t)	Ratio
Poultry	10	17	1.7
Pig	10	23	2.3
Beef	10	30	3.0
Lamb	10	22	2.2
Eggs	6.2	12	1.9
Milk	3	2.7	0.9

Source: J. Woods *et al.*, "Energy and the Food System", Phil. Trans. R. Soc. B, 365 (2010).

4.5   The Arrow of Time

Why do we remember the past and not the future?

This question may appear strange since it is intuitively obvious that time "flows forward". Thus you might be surprised to learn that physics does not distinguish future from past.[5] The equations that describe the universe are just as happy to run the other way. Presumably these equations describe everything, and we are part of everything, so how is it that time appears to flow into the future?

This question was first raised by British astronomer and science popularizer Sir Arthur Eddington (1882–1944) in 1927.[6] Eddington proposed a solution that relies on entropy that is widely accepted today.

Eddington's idea is that time's arrow arises due to directionality implied by the second law of thermodynamics; namely, the future is the direction in which entropy increases. Let's consider a simple example to illustrate this. Think again of the perfume bottle in a room. We associate the future with the molecules of perfume diffusing throughout the room. Why do we not think of running the experiment in reverse and imagine the molecules drifting into the bottle? Nothing in the microscopic laws of physics says that this cannot happen, but the statistical laws of many particles says that it is effectively impossible.

To see this, we simplify heavily and imagine that there are only ten perfume particles. We will ignore all properties of the particles (speed, orientation, spin, identity, ...) except position. We also imagine that space is divided up into small cubes just large enough to hold a single particle. We take the perfume bottle to be only 10 cubes in size, and finally, we imagine a room that is 99 times bigger than the bottle, so that there are 1000 cubes available to the perfume particles once the bottle is open.

We now seek to determine the number of ways in which the perfume particles can be placed in the bottle and compare that to the number of ways they can be placed in the room. Place one particle into the bottle at a time: the first particle has 10 possible cubes to go into; the second then has 9; and so on until the tenth, which must go into the final cube. Thus the total number of ways to place the particles in the bottle is $10 \cdot 9 \cdot 8 \cdots 2 \cdot 1 = 3.6 \cdot 10^6$. Compare this to the number of ways to place them in the room, which is $1000 \cdot 999 \cdot 998 \ldots \cdot 991 = 9.6 \cdot 10^{29}$. This is *immensely* larger than the first. If you placed the particles into the room at random once per second, it would take $6 \cdot 10^{16}$ years to get them all in the bottle (on average). For comparison, the age of the universe is a scant $1.4 \cdot 10^{10}$ years! Thus it is essentially impossible to "run the film in reverse" and perceive perfume molecules assembling themselves in a bottle. Time has a direction because of this practical reality concerning large numbers of particles.

---

[5] Almost. There are rare decays of certain obscure particles that can distinguish past from future.

[6] Who you already know about if you are reading this book backwards.

REVIEW

Important terminology:

entropy   [pg. 90]

joule, Calorie, electron volt   [pg. 88]

kinetic and potential energy, $KE = \frac{1}{2}mv^2$ [pg. 85]

watt   [pg. 89]

Important concepts:

Energy is conserved.

Entropy increases/disorder increases/the quality of energy decreases.

The flow of energy can be followed through complex systems such as the body, the environment, and the economy.

The source of all energy can be traced to the Big Bang.

A person is a 100 W lightbulb.

Food and oil are energy-rich.

Cheap energy makes us rich.

FURTHER READING

David MacKay, *Sustainable Energy – without the hot air*, UIT Cambridge Ltd, 2009.

V. Smil, *Energy in Nature and Society*, MIT Press, 2008.

EXERCISES

1. Conservation of Energy.

   A brick slides across a table with a kinetic energy of 1200J. It comes to a stop after 6s when its kinetic energy is 0J.

   (a) What happened to conservation of energy?

   (b) What is the average power during this 6s?

2. Air Conditioning.

   Estimate the cost of air conditioning the Consol Energy Center for 3 h. In Pittsburgh electricity sells for 10¢/kWh.

3. Baseball.

   Compute the energy of a baseball going at 100 mph. Take the weight of the baseball to be 0.141 kg.

4. 97 W People.

   We established that a person uses energy at the same rate a 97 W light bulb does. A light bulb is too hot to touch. Why is this not true of a person?

5. Proton and Neutron.

   Compute the mass of a proton in kg if its mass is $0.93827\,\text{GeV}/c^2$.

   Compute the mass difference between the neutron and proton in kg. Take the neutron mass to be $0.93956\,\text{GeV}/c^2$.

6. You turn on a 60 W light, which means the light "uses" 60 J of energy every second. Where does this energy go? Can that energy be used to power other things?

7. In the 1990s NASA spent millions of dollars experimenting with something called the "Podkletnov gravity shield" that was reputed to lower the force of gravity in its vicinity. Construct a proof that such a shield cannot exist. Hint: think of a way to use the shield to make a perpetual motion machine.

8. The manufacturers of an exercise bicycle claim that you can lose ½ kg of fat per day by vigorously exercising for 2 h per day on their machine. (a) How many kcal are supplied by the metabolization of 0.5 kg of fat? (b) Calculate the equivalent power. (c) What is unreasonable about the results?

9. Iron Man.

   Iron man can fly to a height of 1 km in 10 s. (a) If he and his suit weigh 400 kg, what power must his suit be able to provide? (b) How much gasoline would he need to carry to power an hour of flight?

10. The world currently uses about 16TW of power. Doesn't this heat up the atmosphere? How much?

11. One could argue that the Green Revolution has permitted one billion extra people to be alive today, at the expense of vast reserves of fossil fuel. Given that fossil fuel use carries a terrible environmental price and that it must run out eventually, was this wise?

12. What fraction of global energy does the USA use?

Electricity, Magnetism, and Light

"I have at last succeeded in illuminating a magnetic curve or line of force, and in magnetizing a ray of light."

— Michael Faraday

"One cannot escape the feeling that these mathematical formulae have an independent existence and an intelligence of their own, that they are wiser than we are, wiser even than their discoverers, that we get more out of them than was originally put into them."

— Heinrich Rudolf Hertz

At a simple level one can think of our world as a complex swirl of four elementary constituents: the electron, proton, neutron, and photon. You are familiar with the first three as the building blocks of atoms: neutrons and protons make up the nucleus (the subject of the next chapter), and electrons orbit the nucleus to make an atom. Finally, the photon is another way to think of light. But a simple list of ingredients is not enough to give understanding. This chapter will sketch the modern interpretation of light. We will see how the theories of James Clerk Maxwell resolved 2500 years of controversy and set the stage for the quantum revolution. Along the way we will explore the important concepts of *force* and *field*.

Maxwell's theory describes and unifies magnetism and electricity – we therefore begin with a historical introduction to both of these phenomena.

5.1 Magnetism and Force

The exploration of magnetism has a long, but inconsistent, history. It was known to the first philosopher, Thales of Miletus (624–546 BC), who wrote on the peculiar properties of magnetic lodestone.[1] The subject appears to have been dropped

---

[1]This stone was found near Miletus at a place called Magnesia; hence the name.

© Springer International Publishing Switzerland 2016
E.S. Swanson, *Science and Society*,
DOI 10.1007/978-3-319-21987-5_5

for 1700 years and then picked up again by French scholar Pierre Pelerin de Maricourt (fl. 1269), who conducted some experiments and wrote a book on the subject.

Magnetism remained a curiosity until 1600 when English natural philosopher William Gilbert (1544–1603) published *De Magnete*.[2] Gilbert's experiments with a small model of the globe led him to conclude (correctly) that the Earth is magnetic, thereby explaining how compasses work. He also made the inspired guess that the center of the Earth was iron. Lest we think that Gilbert had a magical path to knowledge, his guesses about the tides were as misguided as those of Galileo (Sect. 3.2.1):

> "Subterranean spirits and humors, rising in sympathy with the moon,
> cause the sea also to rise and flow to the shores and up rivers."

Gilbert's main contribution to science was not his explorations of magnetism, but the way he *conceptualized* magnetism. Until Gilbert's day, people understood that things moved, metals rusted, and plants rotted, but it was not understood why. It was thought that all things were alive in some way or that things changed because they were seeking their natural place. Gilbert's radical idea was that things moved because a magnetic force makes them move:

> "The force which emanates from the moon reaches to the earth, and,
> in like manner, the magnetic virtue of the earth pervades the region of
> the moon: both correspond and conspire by the joint action of both,
> according to a proportion and conformity of motions, but the earth
> has more effect in consequence of its superior mass; the earth attracts
> and repels the moon, and the moon within certain limits, the earth;
> not so as to make the bodies come together, as magnetic bodies do,
> but so that they may go on in a continuous course."

This novel idea was soon taken up by other natural philosophers. Johannes Kepler, for example, viewed the sun as a symbol of God the Father and a source of motive force in the solar system. This motive force was postulated in analogy to Gilbert's magnetic force. The most famous extension was due to Newton, who guessed that a universal gravitational force governed earthly gravity and the motion of the planets. From this point it was a natural extension to regard all mo-

---

[2]Gilbert was personal physician to Queen Elizabeth I. He died from the plague shortly after his treatise was published.

tion (for example a chemical reaction) as being due to the application of unknown forces.[3] And indeed, it was not long before the electric force was discovered and characterized.

5.2 Electricity

The earliest explorations of the electronic structure of matter dealt with its most familiar form: static electricity. In 1732, London pensioner Stephen Gray discovered that some materials were capable of transmitting charge (conductors), while other were not (insulators). French chemist Charles-François de Cisternay Du Fay observed (1773) that there are two types of electricity, which he called vitreous and resinous, and that likes attract and unlikes repel. In 1745 two Dutchmen, Pieter van Musschenbroek and Andreas Cunaeus, invented the Leyden jar, which is a device that can store electric charge built up by rubbing (Fig. 5.1).

Figure 5.1: A Leyden jar being charged by a spinning pith ball.

Leyden jars were used by Ben Franklin in his famous experiments in the 1740s. Franklin named the battery, renamed the two types of electricity as positive and negative, invented the lightning rod, proved that lightning is electric, and claimed that the amount of electricity does not change. The latter statement is an example of a conservation law (see Chap. 1). The modern form of this law is called "the conservation of electric charge", which states that the total charge in a closed system cannot change.

In 1780 Galvani and Volta disagreed on what makes a frog's leg twitch, as we have seen (Sect. 3.3.3) and Volta constructed his battery.

Quantitative studies of static electricity started with French military engineer Charles-Augustin de Coulomb (1736–1806). Coulomb developed a sensitive torsion balance that allowed him to measure the tiny forces between charged balls. His results were published in a series of memoirs and established "Coulomb's

---

[3]"For many things lead me to have a suspicion that all phenomena may depend on certain forces by which the particles of bodies, by causes not yet known, either are impelled toward one another and cohere in regular figures, or are repelled from one another and recede." – Newton

law", namely that oppositely charged spheres attract each other with a force that is proportional to the product of the quantities of charge and that the strength of the force between the balls drops as the distance between them squared.

Coulomb's law is strikingly similar to Newton's law:

$$F = \frac{1}{4\pi\epsilon_0}\frac{q_1 q_2}{r^2} \text{ [Coulomb]} \qquad F = G\frac{m_1 m_2}{r^2} \text{ [Newton]}. \qquad (5.1)$$

The distance between two bodies is called r in these equations. The factors in front are both constants: G is Newton's constant and $1/(4\pi\epsilon_0)$ is a way to write Coulomb's constant that uses the permittivity of free space, ϵ_0, which will feature in several important formulas discussed below. The final ingredient in Coulomb's law is the magnitude of the charges of the electric bodies, denoted q_1 and q_2.

5.3 Faraday and Fields

The exploration of electricity and magnetism entered a heroic era in the nineteenth century. It started by accident during a lecture given by Danish professor of physics, Hans Christian Ørsted (1777–1851).[4] While making a demonstration, Ørsted noticed that a compass needle deflected when an electric current was allowed to run through a nearby wire, thereby establishing a relationship between electricity and magnetism.

French physicist André-Marie Ampère (1775–1836) was aware of Ørsted's finding, and sought to extend it. He felt that if a current in a wire exerted a magnetic force on a compass needle then two wires should also interact magnetically. In a series of ingenious experiments he showed that this interaction followed simple mathematical laws that were strangely similar to Coulomb's law for electric charge: parallel currents attract, anti-parallel currents repel and the force between two long straight parallel currents is inversely proportional to the distance between them and proportional to the intensity of the current flowing in each.

As important as all these contributions to the science of electromagnetism were, they pale compared to those of Michael Faraday (1791–1867). Faraday was born in Newington Butts, England to modest circumstances. Almost entirely self-educated, he apprenticed to a bookbinder at age 14. Seven years later, Faraday attended a lecture by famous chemist Sir Humphrey Davy of the Royal Society.[5] Faraday sent Davy an extensive set of notes based on these lectures, and Davy was sufficiently impressed to hire him as a secretary. From this point his natural genius enforced a steady rise in status and position. Eventually he was appointed the first Fullerian Professor of Chemistry at the Royal Institution of Great Britain, but

---

[4]21 April 1820, at the University of Copenhagen.

[5]More fully, the "Royal Society of London for improving Natural Knowledge", founded in 1660 by royal charter.

declined a knighthood, the presidency of the Royal Society, and burial at West-
minster Abbey (Fig. 5.2).

Figure 5.2: Michael Faraday (1842).

Faraday's experimental studies covered a stunning range of topics, chiefly in
chemistry, metallurgy, optics, and electromagnetism. He also made contributions
to environmental science, mining safety and education reform.

Two of Faraday's accomplishments in electromagnetism were the construction
of the first electric motor and the first dynamo.[6] In 1831 Faraday discovered that
a moving magnet could induce an electrical current in a nearby wire. He correctly
interpreted this by claiming that a time-varying magnetic field (discussed next)
induces an electric field. This is called *Faraday's law of induction*, and was to
play a crucial role in the revolution to come.

Although Faraday knew no mathematics, arguably his most important contribution was theoretical. To understand what it was, we need to return to some of
Faraday's experiments with magnets. An illustration of a traditional elementary
school experiment with a bar magnet and iron filings is shown in Fig. 5.3. As
can be seen the filings tend to align themselves in a distinctive pattern. Faraday
imagined that each bit of metal was responding to something magnetic that exists
everywhere in space. He called this a *magnetic field*.

In general a field is a quantity that depends on space and time. For example
temperature is described by a field because temperature depends on where you
are, and it changes with time. Mathematically one writes this as $T(x, y, z, t)$.
Some fields also have a direction associated with them; for example wind velocity
depends on space and time, and has a strength and a direction. This is called
a *vector field*. Since we indicate vector quantities by drawing little arrows over

---

[6]A dynamo is an inverse motor, producing electricity from rotational motion.

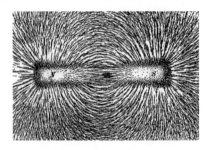

Figure 5.3: Iron filings in the magnetic field of a bar magnet.

them, the wind vector field can be denoted $\vec{v}(x, y, z, t)$. Sometimes one calls fields that do not depend on direction, like temperature, **scalar fields** to emphasize the difference from vector fields.

Faraday also suggested that a charged ball (or an electron or proton) creates an **electric field** around itself. Both the electric and magnetic fields are vector fields because they point in a certain direction at every location. When another charged ball is brought near to the first it responds according to Coulomb's law by being repelled or attracted. Faraday's insight is that this response is not due to a remote effect of the first ball on the second, but is because the first ball makes a field, and the second responds to that field at its location. In a similar way, a massive body like the sun creates a **gravitational field** that acts on the Earth and causes its motion.

Faraday did not think of his fields as conveniences or as imaginary, but rather he envisioned them as real physical entities. He wrote:

> "In the numerous cases of force acting at a distance, the philosopher had gradually learned that it is by no means sufficient to rest satisfied with the mere fact, and has therefore directed his attention to the manner in which the force is transmitted across the intervening space."

and

> "When we turn to radiation phenomena, then we obtain the highest proof, that though nothing ponderable passes, yet the lines of force have a physical existence independent, in a manner, of the body radiating, or of the body receiving the rays."

Source: M. Faraday, "On the physical character of the lines of magnetic force", Philosophical Magazine, June 1852, pg 407.

| Fields are physical entities. |

The introduction of the field concept was a momentous occurrence in the history of science. Faraday's idea resolved a 150 year old conflict between two titans of the previous century: Isaac Newton and German mathematician Gottfried Wilhelm von Leibniz (1646–1716). At issue was the nature of Newton's universal gravity, which acted across enormous distances. How could things so far apart affect each other? Leibniz and others regarded this claim as unscientific since it resurrected old ideas of the vacuum and *action at a distance*, which have a tinge of the occult.

Newton felt uncomfortable with his theory for the same reasons, but had no recourse because his equations implied action at a distance. In desperation he speculated that gravity was a tendency to move to rarer ether – or was simply produced by God. To Leibniz he rather testily replied that, "Gravity really does exist and act according to the laws which we have explained".

If fields are real, as Faraday says, then the remote aspect of Newton's theory can be dealt away. But this leaves another problem, namely Newton's equations also implied that the action of the sun on the Earth was *instantaneous* – it took no time for the gravitational effect of the sun to make its impact felt on the Earth. Again, this had a tinge of the occult. Unfortunately, Faraday's field model does not resolve the issue because it was not known if the field itself was created instantaneously. The solution to this problem was 13 years and another towering genius of the nineteenth century away.[7]

5.4   Maxwell and Light

Unlike Faraday, James Clerk Maxwell (1831–1879) was born in Edinburgh, Scotland to comfortable circumstances (Fig. 5.4). His father was a barrister and a member of the aristocratic Clerk family. James was sent to a boarding school at a young age, where his provincial accent earned him the nickname, "Daftie". In spite of this, he published his first scientific paper at the age of 14. He is widely regarded as one of the most important scientists of the nineteenth century, and the impact of his work is ranked with that of Isaac Newton and Albert Einstein.

Maxwell worked on a variety of topics in mathematical physics in his career. His creativity was such that he often invented fields while solving problems. Maxwell correctly predicted that the stability of the rings of Saturn implies that they are made up of particles (rather than a liquid or a solid). He also contributed

---

[7]Faraday was acutely aware of this issue, "There is one question in relation to gravity, which, if we could ascertain or touch it, would greatly enlighten us. It is, whether gravitation requires time. If it did, it would show undeniably that a physical agency existed in the course of the line of force." *Experimental Researches in Electricity*, 1852.

Figure 5.4: James Clerk Maxwell.

to, or invented, color theory, control theory, kinetic theory, and thermodynamics. Unfortunately this prodigious work was cut short because Maxwell died of abdominal cancer at age 48.

Maxwell's crowning achievement was in correcting and unifying the various laws that described electricity and magnetism in a set of four equations, called, unsurprisingly, **Maxwell's equations**. In case you want to put them on a t-shirt, here they are:

$$\nabla \cdot \vec{E} = \frac{1}{\epsilon_0}\rho$$

$$\nabla \cdot \vec{M} = 0$$

$$\nabla \times \vec{E} = -\frac{\partial \vec{M}}{\partial t}$$

$$\nabla \times \vec{M} = \mu_0\epsilon_0\frac{\partial \vec{E}}{\partial t} + \mu_0\vec{J}$$

The electric and magnetic fields are denoted $\vec{E}$ and $\vec{M}$ (remember that they are vector fields and hence have arrows). The charge, either at rest or flowing, is represented by the symbols ρ and $\vec{J}$. The last elements in Maxwell's equations are ϵ_0, which is the **permittivity** of free space, and μ_0, which is the **permeability** of free space. These are constants that are required to make the units of the equations work out, and measure properties of the vacuum. The upsidedown triangles on the left of Maxwell's equations indicate rates of change in space, while the symbols $\partial/\partial t$ on the right indicate rates of change in time.

The first of these equations is the mathematical transcription of Coulomb's law in terms of fields. The second is a condition on the magnetic field.[8] The third is

---

[8]The zero on the right hand side of this equation is significant. It implies that a magnetic analogue of electric charge, called a **magnetic monopole**, does not exist.

Faraday's law of induction, and the fourth is Ampère's law with the addition of an extra term. Maxwell guessed that this term should exist because the third equation involves the time rate of change of the magnetic field, so perhaps a changing electric field can produce a magnetic field. The guess turned out to be correct and momentous, for it led Maxwell to the solution of a 2500 year old problem.

5.5 Vision and Light

At issue was the nature of light, which roiled the (rather small) world of natural philosophers since the ancient Greeks. We will briefly review the problems that arose and then discuss Maxwell's resolution in this section.

5.5.1 Theories of Vision

For millennia light was not thought of as a thing, but rather as a property of vision. The Greek philosopher Empedocles (490–430 BC) was the first to devise a comprehensive theory of light and vision. Empedocles postulated that we see objects because light streams out of our eyes and illuminates the objects. This idea, called the *extramission theory* of vision, became the working assumption for much of the later developments in optics and vision. The problem, of course, is that the model has reversed the direction of causality. Empedocles appears to imagine that things are visible because we make them so with our eyes. But this is backwards, we see things because they reflect light into our eyes.

The Greek atomists, led by Democritus, also ventured a theory of vision in which sight is attributed to the reception of a thin film of atoms, called a *simulacrum*, which is emitted from the surface of all objects. Presumably the simulacrum is only emitted under illumination, otherwise we would be able to see in the dark.

The Aristotelian concept of light and vision was equally imaginative and wrong. According to Plato (in the *Timaeus*), vision is the result of fire issuing from the observer's eye whence it coalesces with sunlight to form a medium. This medium stretches from the visible object back to the eye, permitting its perception. Aristotle's view was based on his potentiality theory and was more subtle than that of his teacher. He held that a visible object produces an alteration of the transparent medium by changing it from a state of potentiality into one of actuality. The medium then transmits this alteration to the eye, where it is subsequently perceived. Light itself is simply the state of the transparent medium. Notice that Aristotle has broken with his teacher and joined the atomists by constructing an *intromission theory*, wherein light carries information *into* the eye, rather than out from the eye.

Euclid (323–283 BC) and Ptolemy (90–168) both wrote books called *Optics* that dealt with geometric aspects of vision. Ptolemy attempted to combine Euclid's geometrical approach with the physical and psychological aspects of vision.

He correctly stated that the angle of incidence equals the angle of reflection for reflected light and developed a large set of theorems on the properties of reflected images. Ptolemy also examined the phenomenon of *refraction*, which is the bending of light that occurs when it passes from one transparent medium to a different one (Fig. 5.5). Here Ptolemy took a modern point of view and experimentally tested the effect with the goal of determining a mathematical relationship between the angle of incidence and the angle of refraction. Although it was evident to Ptolemy that such a relationship existed, he was not able to guess the formula.

Figure 5.5: The refraction of light through a glass block.

The Arabic natural philosopher Alhazen (965–1040)[9] made an important advance by demolishing the extramission theory of vision that had survived for 1500 years. How is it, Alhazen argued, that one can perceive the immensely distant stars the instant one's eye open? This either requires extramission visual rays that travel instantaneously – something that Alhazen rejected on good grounds – or an intromission theory. Furthermore, as is immediately clear to anyone who has looked at the sun for too long, light affects the eye, it is not the eye that affects the sun. Alhazen backed up his assertions with careful experimentation, demonstrating that light is generated by luminous objects and it is this light that is received into the eye. Thus "the extramission of [visual] rays is superfluous and useless."[10]

Some of Alhazen's demonstrations were made with a *camera obscura*, familiar to school children as the pinhole camera. By passing light from various sources through a small hole and into a darkened chamber and onto a screen, he was able to show that light travels in straight lines and that images do not get confused upon passing through the hole. Thus Alhazen established a direct model of the eye, with the lens acting as the pinhole, the eye acting as the chamber, and the retina acting as the screen. In fact the words *lens*, *retina*, and *cornea* come from Latin via medieval translations of Alhazen's *Book of Optics*. It is perhaps not

---

[9]Alhazen is the latinized version of Abu Ali al-Hasan ibn al-Hasan ibn al-Haytham

[10]Source: A.M. Smith, *Review of A. I. Sabra, The Optics of Ibn al-Haytham. Books I, II, III: On Direct Vision*, The British Journal for the History of Science **25**, 358 (1992).

surprising that Alhazen also studied the anatomy of the eye and that he decisively distinguished the physics of optics from the psychology of vision.

In studying refraction, Alhazen made another prescient claim by correctly concluding that the phenomenon happens because light slows down as it enters a dense medium. Thus light must move with a finite speed.[11]

Theories of vision continued to be improved over the centuries and reached a high level of perfection in the work of Hermann von Helmholtz (1821–1894), who developed and tested theories of color vision and depth and motion perception and championed the importance of the brain in forming these perceptions.

5.5.2 The Particle-Wave Controversy

By the 1600s, the controversy over extramission and intromission had withered and was replaced with a new one concerning the ultimate nature of light. The issue was in fact ancient: atomists like Democritus thought that light consisted of invisible particles while Aristotle and others thought of it as undulations (or something similar) in an invisible medium.

French philosopher René Descartes (1596–1650) earned fame by seeking to replace the Aristotelian system with a theory of matter built from pure logic.[12] Part of his universal model postulated that light was wave-like and, in agreement with Alhazen, that refraction occurred because the speed of light changes in different substances.

A countryman, natural philosopher and priest Pierre Gassendi (1592–1655), argued against Descartes. Gassendi endeavored to reconcile Democritean atomism with Christianity and maintained that light was particle-like. Isaac Newton had read Gassendi and preferred his views to those of Descartes (whose general philosophy Newton disliked). He argued that light had to consist of "subtle" particles because it did not bend upon hitting an obstacle, unlike water waves. Refraction was explained by the particles accelerating (due to gravity – which is wrong) when entering a medium. The reflection of light was explained by particles bouncing off of reflecting substances, much like billiard balls bouncing off of a cushion.

English natural philosopher Robert Hooke (1635–1703), and sometimes bitter rival of Newton, took up the wave side of the argument, and correctly suggested (1672) that the vibrations of light were perpendicular to the direction of propagation.[13] This view was most famously taken up by Dutchman, Christiaan Huygens (1629–1695) and developed into a functional mathematical theory. Huygens held that the speed of light should be finite and that it was emitted in spherical waves

---

[11]Source: H. Salih, M. Al-Amri and M. El Gomati, *The Miracle of Light*, A World of Science, **3**, 2 (2005).

[12]He failed dramatically, but the idea of working logically and with the aid of observation helped launch the Scientific Revolution.

[13]Thus light is a transverse wave, a topic we will take up again in the next section.

from sources. Like many before him, Huygens assumed that light was a wave *in* something, which he called the ***luminiferous ether***.

Despite these successes, Newton's view held sway because of his formidable reputation. The tide finally began to reverse in 1803 when English scientist Thomas Young (1773–1829) performed his famous ***double slit experiment***. Young first established that water and sound waves can interfere with each other to produce a characteristic pattern, called ***diffraction*** (see Fig. 5.6). He then showed that precisely the same phenomenon happened, but at much smaller scales, with light. His experimental results permitted him to compute the wavelength of red and violet light, giving him values that agree with modern ones.

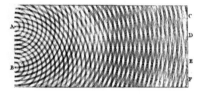

Figure 5.6: Young's depiction of diffraction (1803). A wave source lies to the left and passes waves through two holes labelled A and B. Wave interference is observed at the points labelled C, D, E, F.

The battle was finally won in 1817 when Augustin-Jean Fresnel (1788–1827) submitted an essay for a prize offered by the French Academy of Sciences. Fresnel's essay used wave theory to explain diffraction, which did not sit well with Siméon Denis Poisson (1781–1840), who was both a member of the examining committee and a particle theorist. Poisson thought he had found a flaw in young Fresnel's argument when he realized that it implied that there should be a bright spot in the shadow of a circular obstacle that blocks the light from a point source. This nonsensical prediction (in the particle theory) convinced Poisson that Fresnel was wrong, but the head of the examining committee thought the idea worth testing and found the spot.[14] Fresnel won the prize and wave theory won the day.

This is not the end of the story: currently it is believed that light is *both* wave-like and particle-like, making Newton and Hooke simultaneously right and wrong. Exactly how this works is described in Sect. 6.4.

5.5.3   Maxwell's Radiation

By the 1860s the idea that light was an undulation in the ether was firmly established. The questions remained, what kind of undulation, and what was the ether?

Maxwell had an answer to the first question in 1862. Recall that a changing magnetic field produces an electric field and that a changing electric field produces

---

[14]Now named in honor of the committee head as the ***Arago spot***.

a magnetic field. Thus if one creates a changing electric field (say by wiggling a charged ball) then it is possible for a magnetic field to be created, which in turn creates an electric field, and so on. In this way the entwined electric and magnetic fields can bootstrap their way across space (see Fig. 5.7). Maxwell realized that his equations described exactly this situation, and furthermore, they predicted that the speed of this disturbance was given by

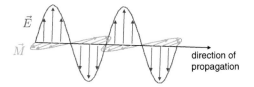

Figure 5.7: A model of light as electromagnetic radiation.

$$c = \frac{1}{\sqrt{\epsilon_0 \mu_0}}. \tag{5.2}$$

Upon working out the numbers, Maxwell obtained a speed of $3.107 \cdot 10^8$ m/s, a number remarkably close to the speed of light (the modern value is $2.9979 \cdot 10^8$ m/s). This was more than mere coincidence:[15]

> "The agreement of the results seems to show that light and magnetism are affections of the same substance, and that light is an electromagnetic disturbance propagated through the field according to electromagnetic laws."
>
> — J.C. Maxwell

Thus we say that light is *electromagnetic radiation*.

Maxwell had no answer for the second question: the nature of the ether, but firmly believed that it existed. His stubbornness must be understood in the context of long and entrenched thinking about the ether. Yet Maxwell's equations do not require the ether and the bootstrapping mechanism we have described also does not require it.[16] The concept of the ether was finally laid to rest by a combined effort of Armand Hippolyte Louis Fizeau (1819–1896), American physicists Albert Michelson (1852–1931) and Edward Morley (1838–1923), and Albert Einstein.

The beginning of the end for ether was the *Fizeau water experiment* of 1851, where Fizeau showed that the speed of light in moving water implied that the ether was partially "dragged" by the water. In 1887, Michelson and Morley showed that

---

[15] In fact Faraday had guessed that light was electromagnetic in character decades before.

[16] Look back at the quotation from Hertz that opened this chapter, as you will understand it better now.

the speed of light does not change when it is measured moving with the Earth or a right angles to the Earth; thus the ether is *not* dragged by the Earth. Finally, Einstein reconciled both results with his ***Theory of Relativity*** in 1905 and put the ether out of its misery.

The development of a model of light did not stop with Maxwell's breakthrough. Sixty years later it was realized that the microscopic world is quantum; by 1950 a quantum version of Maxwell's equations had been developed. In the 1960s this theory was incorporated into a larger one that included three of the four known forces, called the ***Standard Model***. Today people continue to explore ways to extend this theory. The next chapter will describe some of these revolutionary ideas and apply them to understanding common experience. But first we will describe light in more detail.

5.6 Measuring Light

Light is the best understood phenomenon in the universe because it is so simple. In fact it can be described with just three numbers, ***intensity***, ***frequency***, and ***polarization***. The intensity of light is a measure of brightness. Since light is a wave, it oscillates at a certain frequency (or mixtures of frequencies), which is measured in "cycles per second". The unit for frequency is thus $1/s$, which is called a "hertz" (abbreviated as "Hz"). The speed of light is always c in a vacuum, independent of the frequency. This means that the ***wavelength*** of light is simply related to the frequency by the formula

$$\lambda = \frac{c}{f}. \tag{5.3}$$

The traditional notation of λ for wavelength, c for the speed of light, and f for frequency has been used in this equation.

The polarization of light is defined by the plane in which the electric field oscillates. Take a look at Fig. 5.7 again; if the magnetic field is along the x-axis, then the electric field is along the z-axis, and the light propagates along the (positive) y-axis. We say the polarization is in the $y - z$ plane. Most natural light is unpolarized, which means that the atoms that create the light do so with random orientation of the electric field so that the produced light is continuously (and quickly) changing its orientation. Some transparent materials will only let light that oscillates in a particular direction pass. The polaroid film that coats expensive sunglasses is one such material. We will learn why polaroid sunglasses are useful in the following chapter.

Notice that the oscillations of the electric and magnetic fields happen in directions that are perpendicular to the direction of propagation. This is a general feature of electromagnetic radiation, and we say that light is a ***transverse*** wave. This is distinct from sound waves where the oscillations occur along the direction of propagation. We thus refer to sound as a ***longitudinal*** wave. The rolling motion of water waves is quite complex and in fact they are both longitudinal and transverse.

As we will discuss in the next chapter, at very short distances light is best described as a quantum phenomenon. In this case light comes in packets of energy called **photons**. Einstein and German physicist Max Planck (1858–1947) wrote an equation relating the energy of a photon with its frequency:

$$E = hf. \tag{5.4}$$

You can guess what E and f stand for. The new quantity is called **Planck's constant** and is given by

$$
\begin{aligned}
h &= 4.135 \cdot 10^{-15} \,\text{eV} \cdot \text{s} \\
&= 6.626 \cdot 10^{-34} \,\text{J} \cdot \text{s}. \tag{5.5}
\end{aligned}
$$

The units of Planck's constant are `energy·time`, which we write as eV·s or J·s. Planck's constant will make another appearance in mathematical models of atomic structure (Chap. 6).

These equations imply that light can be characterized equally well by frequency, wavelength, or energy since they can all be freely interchanged. The totality of frequencies available (which is infinite) is called the **spectrum** of electromagnetic radiation. The spectrum is shown in Fig. 5.8, labelled by frequency, energy, and wavelength. The figure also displays the "absorption coefficient" of light, which we will discuss in Sect. 6.8.1.

Some of the spectrum has names that you may have heard of:

- A **gamma ray** is high energy, small wavelength light. For this reason gamma rays are dangerous to health. We will discuss them more in the next two chapters.

- **X-rays** refer to light with an energy of about 1000 electron volts (100–100,000 eV) and are familiar to us by their use in dentistry and medicine.

- **Ultraviolet radiation**, often called *UV*, has an energy just above that of the visible portion of the spectrum.

- We are also familiar with **infrared radiation** via infrared cameras that can "see in the dark".

- **Thermal radiation** is electromagnetic radiation that is emitted due to the thermal motion of charged particles. It can occur anywhere in the spectrum, depending on the temperature of the emitting material.

- **Microwave** ovens emit electromagnetic radiation at a frequency that is tuned to excite motion in water molecules, and thereby heat up food. The microwave portion of the spectrum is also used in wifi and cell phone communication.

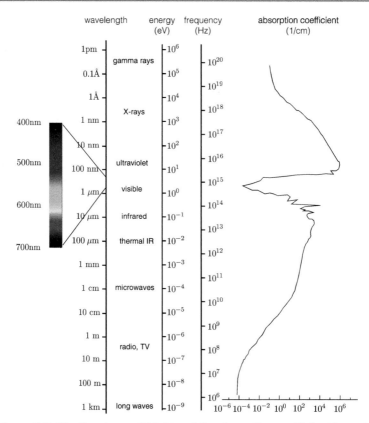

Figure 5.8: The Spectrum of Light and the absorption coefficient in water.

- **Radio waves** are typically of wavelength 1–10 m, with FM radio covering 87.9 to 108 MHz and AM radio broadcasts in the 153–270 kHZ range, while short wave radio uses the 2.3–26.1 MHz band.

- Of course the **visible** part of the spectrum is that (tiny) portion that the eye can discern. It is traditionally split into seven parts (red, orange, yellow, green, blue, indigo, violet), but of course the frequencies form a continuum and so the visible spectrum cannot be described with a finite number of labels. It is traditional to describe the rainbow in terms of seven colors only because Newton named them, and he deemed seven a lucky number.

5.7   Spectral Lines

By the late nineteenth century it was possible to measure the spectrum of light that was produced by a light source. For example if a pure blue light was analyzed one would find a single "line" at a frequency of 650 THz. A line like this is called an *emission line*. Alternatively, consider a light source that emits at all frequencies that shines through a gas and is analyzed. In this case some of the continuous spectrum that is expected will not appear – lines will be missing. One then speaks of *absorption lines*, with the idea that the gas has "absorbed" light at certain frequencies.

It turns out that different gases absorb different parts of the spectrum, or have different *absorption spectra*. Thus it became possible to identify the chemical composition of gases by measuring their absorption spectra. Extensive experimental effort soon revealed that some spectra were remarkably simple. For example, the spectrum for hydrogen is simple enough that a formula for the absorbed frequencies was guessed by Johannes Rydberg in 1888:

$$f = R \left(\frac{1}{n^2} - \frac{1}{m^2} \right), \tag{5.6}$$

where n and m are integers with $n < m$, and R is called the *Rydberg constant*

$$R = 3.28 \cdot 10^{15} \text{ Hz.} \tag{5.7}$$

The realization that spectra are linked to chemical properties of gases immediately proved useful. For example, examining the spectrum produced by the sun revealed absorption lines that agreed with those of hydrogen, showing that the sun is largely a ball of hot hydrogen. It also revealed lines that had not been observed before, but were subsequently found on Earth and identified as a new element (helium!).

Although light was thoroughly understood by the close of the nineteenth century, the discovery of absorption and emission spectra made it plain that the interaction of light with atoms was mysterious. Resolving this mystery did not take long and ushered in yet another revolution in our conception of nature.

REVIEW

Important terminology:

absorption and emission spectra   [pg. 119]

diffraction   [pg. 114]

light intensity, wavelength, frequency, and polarization   [pg. 116]

photons, $E = hf$ [pg. 117]

reflection   [pg. 112]

refraction   [pg. 112]

scalar and vector fields   [pg. 108]

Important concepts:

Action at a distance.

The luminiferous ether.

Like charges repel, unlike attract.

Coulomb's Law.

Ampère's Law.

Faraday's Law of Induction.

Fields are physical entities.

Intromission and extramission theories of vision.

Light is transverse electromagnetic radiation.

FURTHER READING

L. Campbell and W. Garnett, *The Life of James Clerk Maxwell*, MacMillan and Company, 1882. Available online at
http://www.sonnetsoftware.com/bio/maxbio.pdf.

J. Henry, *The Scientific Revolution and the Origins of Modern Science*, Palgrave, New York, 2002.

M.C. Jacob, *The Scientific Revolution: a brief history with documents*, Bedford/St. Martin's, New York, 2010.

D.C. Lindberg, *The Beginnings of Western Science*, University of Chicago Press, 2008.

S. Schweber, *QED and the Men Who Made It*, Princeton University Press, 1994.

EXERCISES

1. Magnetic Fields.

   Look again at Fig. 5.3. You would think that the magnet should create a field that is smooth, so why do the filings tend to line up with spaces between the lines?

2. Islands in the Sun.

   It is reported that ancient Polynesian explorers could sometimes tell they were near an island before they could see it. Suggest a way that this might work based on considerations in this chapter.

3. Field Speeds.

   If electric and magnetic fields propagate at the speed of light, how fast do you think a gravitational field propagates? These are two of the four forces of nature; what are the other two? Do they have fields? If so, how fast do these fields propagate?

4. Waves.

   Categorize the following waves as transverse, longitudinal, or both.

   (a) light
   (b) sound
   (c) water
   (d) a wiggling string
   (e) a wiggling slinky
   (f) a stadium wave.

5. Fields.

   Categorize the following as scalar fields, vector fields, not fields.

   (a) the velocity of water at every point in a stream
   (b) the direction a compass points on the surface of the Earth
   (c) the speed of a car
   (d) the force of gravity near the Earth
   (e) the pressure throughout the ocean
   (f) the humidity in the atmosphere
   (g) the average height of US men.

6. Atomic Transitions

   Compute the energy, wavelength, and frequency of a photon that is emitted when hydrogen makes a transition from the fifth to the second orbital.

Atom and Light

"I have broken the machine and touched the ghost of matter."

— Ernest Rutherford

We have spent considerable effort learning about light and its properties, but light is most interesting when it interacts with matter. In this chapter we will learn about the structure of the atom and how atoms and molecules interact with light. This necessarily involves the ***quantum*** concept that underpins the microscopic world and revolutionized our understanding of nature. With this information in place we will tackle polarization and scattering of light ("why do sunglasses work?" and "why is the sky blue?"), and examine the longstanding claims that electromagnetic radiation harms the body.

6.1 The Electron

Although the atomic paradigm for matter had existed since at least the days of Democritus, it had gained serious following after John Dalton's (1766–1844) breakthrough in understanding chemical reactions. There was still, however, no idea that an atom could be *made* of things. In fact the natural assumption was that hydrogen was the base atom upon which all others were built.

This idea was definitively demolished in 1897 by Joseph John Thomson (1856–1940). Thomson was a British physicist who had succeeded Lord Rayleigh as head of the Cavendish Laboratory at Cambridge University. The Cavendish was founded by Maxwell in 1874 and was the leading center of subatomic research in the world for 80 years. That it remains a center of excellence in science is evidenced by the 29 Nobel Prizes Cavendish researchers have won.

Thomson's breakthrough experiment was made with a ***Crookes tube*** (the same Crookes mentioned in Sect. 3.2.1), which is a glass tube with an embedded wire that – we now know – emitted electrons when current was passed through it. This is called a ***cathode*** and is shown in Fig. 6.1 as "C". The electrons are attracted to a positively charged plate, labelled "A" in the figure, pass through a gap ("B") into the main body of the tube, and finally they impact a fluorescent screen (seen at the far right in the figure) where the subsequent tiny flashes of light can be seen and recorded (by dark-adjusted and very patient assistants).

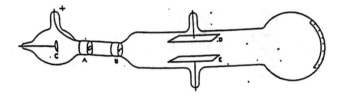

Figure 6.1: Cathode ray apparatus, as drawn by Thomson.

Experiments with Crookes tubes had established that something was being emitted by the cathode, but what? Thomson made three important modifications to the experiment in an attempt to find out. First, he evacuated the tube so that air would not interfere with the free motion of the unknown rays. Second, he placed charged plates ("D" and "E") in the tube to determine if the rays could be deflected by electric fields, and third, he placed the tube in the field of a large magnet to test the effect of magnetic fields on the rays. To his surprise, he found that electric and magnetic fields could both deflect the beam. By carefully balancing the strengths of the electric and magnetic fields and assuming that he was dealing with a beam of particles of mass m and charge q he was able to measure the ratio q/m with great accuracy. He also estimated the particle mass by measuring the heat generated when the particles hit a thermal junction, obtaining a value of about 1/1000 of the mass of hydrogen (the modern value is about 1/1836).

Thomson confirmed that the rays were particle-like by placing an impediment in the beam and observing a sharp shadow. He also showed that hydrogen atoms had one electron each. Finally, Thomson believed that the electrons were coming from traces gases in the Crookes tube. This led him to repeat the experiment with different gases in the tube. The result was the same numerical charge to mass ratio in each case.[1]

Thomson was forced to dramatic conclusions: there had to be a particle, smaller and lighter than hydrogen; this particle was present in all matter, and was the same in all matter. In other words Thomson had discovered a fundamental particle of nature, and it was a building block of atoms.

6.2   Rutherford's Atom

Thomson's discovery immediately raised the question: what is an atom? At the time it was known that they were small, they had a mass of about 10^{-27} kg (for hydrogen), they were neutral (i.e., had no electric charge), and they had electrons

---

[1]Thomson was wrong here, he needed to change the material in the cathode to conclude that the electron was universal.

in them. Thomson thus proposed the ***plum pudding model*** where an atom consists of a small pudding-like body of positive charge with various numbers of electrons spread through the volume (like raisins in a plum pudding). What the "pudding" was remained unknown.

This model was to only last 14 years, when experimental results from a brash New Zealander revolutionized subatomic physics again. The man in question was Ernest Rutherford (1871–1937) (Fig. 6.2), Thomson's student and eventual successor.[2] Rutherford invented and dominated the field of subatomic physics until his death.[3]

Figure 6.2: Baron Rutherford of Nelson.

Among Rutherford's accomplishments were the discovery of radon, the discovery of alpha and beta rays, the discovery of nuclear half lives, the realization that gamma rays were a third kind of radiation (see Sect. 5.6), the proof that alpha rays are the nuclei of helium, the conversion of nitrogen to oxygen, the realization that the nucleus of hydrogen was likely a fundamental particle that he named the ***proton***, and the guess that ***neutrons*** existed in the atomic nucleus. We will be discussing most of these topics in Chap. 8; here we concentrate on Rutherford's model of the atom.

Given this body of work, the obituary from the New York Times does not appear as exaggeration:

"It is given to but few men to achieve immortality, still less to achieve Olympian rank, during their own lifetime. Lord Rutherford achieved

---

[2]Davy once said that his greatest discovery was Faraday. Thomson might well have said the same of Rutherford.

[3]A death caused by an unnecessary delay in surgery for a hernia.

both. In a generation that witnessed one of the greatest revolutions
in the entire history of science he was universally acknowledged as
the leading explorer of the vast infinitely complex universe within the
atom, a universe that he was first to penetrate."

In 1909 Rutherford, Hans Geiger (1882–1945), and (undergraduate!) Ernest
Marsden (1889–1970) decided to test Thomson's plum pudding model by bombarding a thin foil of gold with alpha particles. The team expected that the alpha
particles would undergo slight deflections as they passed through the pudding
atoms that make up the foil. This is indeed what they observed, *except* for rare
(about 1/8000) cases where the alpha particle deflected backwards. The results
stunned Rutherford, who said,

> "It was quite the most incredible event that has ever happened to me
> in my life. It was almost as incredible as if you fired a 15-inch shell
> at a piece of tissue paper and it came back and hit you."

In pondering these remarkable results Rutherford came to the conclusion that
the plum pudding model was "not worth a damn". Detailed calculations convinced him that the alpha particle was bouncing off of a small and dense object,
and not much of anything else. Thus the atom had to consist of a tiny core surrounded by electrons. Rutherford had discovered the **nucleus**. Guided by his results, Rutherford constructed an atomic model that resembles a tiny solar system,
where one imagines that electrons move in orbits around the nucleus, attracted to
it by Coulomb's law instead of Newton's. This is the familiar image that one finds
in elementary schools and in popular culture.

6.3 Blackbody Radiation

While Thomson and Rutherford were making their discoveries another conundrum was being unravelled. At issue was the intensity of light (measured in watts
per square meter) emitted by a **black body**. A black body is a theoretical material
that absorbs radiation perfectly (and then re-emits it). A typical physical model
is an oven with a small hole, although many objects such as planets or stars or
glowing metal pans can be usefully approximated as black bodies.

The radiation emitted by a black body is of special interest because it only depends on the temperature of the black body – the geometry and composition of the
body are irrelevant. You can verify this by connecting two different black bodies
that are at the same temperature. If the blackbody radiation differs, radiation can
flow from one to the other and the other will heat. But this violates the second
law of thermodynamics. Furthermore, if one places a filter that selects a given
frequency of electromagnetic radiation between the black bodies, the same result
holds. These results imply that the *blackbody spectrum is universal*.[4]

---

[4]In other words, it has a fixed dependence on frequency that only changes with temperature.

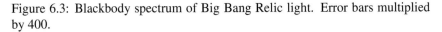

The form of blackbody radiation only depends on temperature.

Although all of this was known by 1860, the measurement of the blackbody spectrum was difficult and was not achieved until 1899 by Otto Lummer (1860–1925) and Ernst Pringsheim (1859–1917). The result was a curve similar to that shown in Fig. 6.3.

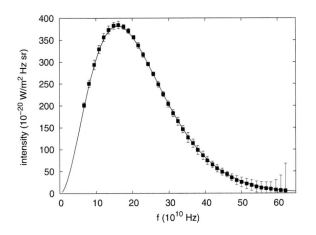

Figure 6.3: Blackbody spectrum of Big Bang Relic light. Error bars multiplied by 400.

The race to explain the curve was won by German physicist Max Planck (1858–1947) in December of 1900.[5] It was accepted that the radiation trapped in a black body oven was in thermal equilibrium with the material of the oven walls. If one assumed that the oven walls were made of microscopic simple oscillators (imagine a mass on a spring wiggling to and fro) then the form of the blackbody radiation curve could *almost* be obtained. The near miss frustrated Planck, but no manipulation could yield a formula that agreed with experiment. Finally, in desperation Planck assumed that the energy of the wall oscillators could only take on certain discrete values. He called this assumption **quantization** in analogy with the Latin word for quantity.

---

[5]Planck endured unspeakable tragedy: his first wife died of tuberculosis, both of his daughters died while giving birth, his oldest son was killed in action in WWI, his second son Erwin was executed by the Gestapo for taking part in a failed assassination of Hilter. Erwin's death finally destroyed Planck's will to live and he died shortly afterward.

We have already seen Planck's quantization assumption in Sect. 5.6: $E = hf$, where h is Planck's constant. His blackbody radiation law that emerged from this assumption is

$$U = \frac{2hf^3}{c^2} \frac{1}{e^{hf/k_B T} - 1},$$ (6.1)

where U is the power per unit area per unit solid angle per unit frequency of the emitted blackbody radiation. You are familiar with all of the symbols in this equation except k_B. This is called **Boltzmann's constant**, named after Austrian physicist Ludwig Boltzmann (1844–1906). The Boltzmann constant relates energy and temperature (you can verify that its units are `energy/temperature`) and often appears in formulas involving thermodynamics. You are probably familiar with the **gas constant** of $PV = nRT$ fame – Boltzmann's constant relates the gas constant to the Avogadro number via $R = k_B N_A$.

We will be interested in the total energy emitted by a blackbody. This can be obtained by summing Planck's formula over all frequencies. The result is called the **Stefan-Boltzmann law** and is given by[6]

$$j = \sigma T^4$$ (6.2)

with

$$\sigma = \frac{2\pi^5 k_B^4}{15c^2 h^3} = 5.67 \cdot 10^{-8} \, \text{W/(m}^2 \, \text{K}^4)$$ (6.3)

The units of j are W/m^2 and the quantity σ is called the **Stefan-Boltzmann constant**. This formula will prove useful many times in the chapters to come since it provides a way to relate the temperature of a (black) body to the power of its thermal radiation.

Finally, let's consider Fig. 6.3 again. The data represent the blackbody radiation intensity of light that was emitted by matter approximately 380,000 years after the universe was created (an event called the Big Bang). The data fit Planck's formula perfectly when a temperature of 2.72548 ± 0.00057 K is assumed for the black body. This temperature matches precisely what is expected if the universe has expanded by a factor of 1000 in the intervening 14.1 billion years and provides a dramatic confirmation of the hypothesis that the universe began some 14.5 billion years ago.

6.4 The Photoelectric Effect

Yet another conundrum was roiling the physics world at the same time that Rutherford and Planck were unraveling their mysteries. In 1887, Heinrich Hertz (1857–1894) had noticed that electrodes emitted sparks more readily when they were illuminated with ultraviolet light. This discovery was followed by a long series of

---

[6]Named for Boltzmann and Slovenian physicist Jožef Stefan (1835–1893).

experiments that demonstrated that many metals and gases emit electrons when light is shone on them. According to Maxwell's equations, this is because the light transfers energy to electrons, which eventually evaporate from the surface of the metal. A higher intensity light should eject electrons in greater numbers and more rapidly than a lower intensity light.

Surprisingly, this is *not* what was observed. Rather, no electrons were emitted at all until a threshold frequency was reached. This is true regardless of the intensity of the light source. The conundrum remained unresolved until 1905 when Albert Einstein boldly guessed that light is a collection of discrete (i.e., quantum) packets, called **photons**. The energy of the packets were related to the frequency of the light that they comprised by the formula that we have seen twice already: $E = hf$. In effect, Einstein had inverted Planck's quantization hypothesis by saying that it was light that was quantized, not matter (i.e., the wall oscillators).

With this guess Einstein was able to show that a certain energy is required to liberate an electron from a metal, otherwise the electron merely absorbs the light and re-emits it. The theory predicted that liberated electrons should have an energy proportional to the frequency of the illuminating light. Although this was soon confirmed, resistance to Einstein's quantum hypothesis was strong because Maxwell's equations had been outstandingly successful.

Eventually it was realized that the Einsteinian and Maxwellian pictures of light could be reconciled if one considered continuous light (also called classical light) to consist of a great many photons. In this way the lumpiness of the light's energy content could not be discerned and a classical description would be accurate. Alternatively, the quantum nature of light is only obvious at very low intensities. This joint nature of light goes by the name of **wave-particle duality** and is regarded as a central feature of light and of nature today.

Blackbody radiation and the photoelectric effect both involve the interaction of light with matter in special circumstances. It is thus no surprise that the discoveries of Planck and Einstein would soon have an impact on the development of the theory of atoms.

6.5 Bohr's Atom

Rutherford's atomic model was not destined to have a long life because it suffered from a serious flaw. Maxwell's equations imply that an accelerated charge will emit electromagnetic radiation – in fact this is how all electromagnetic radiation is created. Radiation carries energy (think about sunlight heating up a road surface), which means that the accelerated charge must lose energy. In the case of Rutherford's atom, the accelerating charge is the electron orbiting the nucleus. The subsequent energy loss forces the electron to quickly (actually within tens of picoseconds!) spiral into the nucleus.

Of course atoms are not disappearing into poofs of smoke so something is wrong.[7] There is no way around this problem, yet the basic structure of the Rutherford model had to be right – some drastic action was required. Fortunately, a young Danish physicist named Niels Bohr (1885–1962) (Fig. 6.4) was willing to take this action. Three years after Rutherford framed his model, Bohr suggested that electrons cannot be any distance from the nucleus but only those distances that obey the relationship

Figure 6.4: Niels Bohr (*left*) and Max Planck (1930). Maxwell's first equation is on the board behind them.

Reproduced with permission from the Niels Bohr Archive.

$$n\frac{h}{2\pi} = \sqrt{\frac{e^2 mr}{4\pi\epsilon_0}}. \tag{6.4}$$

We have seen many of these symbols before. Recall that ϵ_0 appears in Coulomb's law and Maxwell's equations, e is the charge of the electron, r is the radius of the electron's orbit (assumed circular), h is Planck's constant that appeared in Einstein's equation for the energy of a quantum of light, and n is an arbitrary integer denoting the nth permitted electron orbit. Planck's constant had already been associated with quantum (i.e., not everything is allowed) effects by Planck and Einstein. This is yet another quantum effect since Bohr's condition does not allow all possible electron orbits. Because the integer n is associated with a quantization condition it is called a ***quantum number***.

Bohr's quantization condition gives the electron radius from which one can compute the energy of the electron. That energy will depend on which orbit the electron is in, n, and is given by a simple formula:

---

[7]What about the solar system? The same thing happens to it, except in this case planets emit gravitational radiation and the spiral-in time is immensely long.

$$E_n = -\frac{me^4}{8h^2\epsilon_0^2 n^2} = -\frac{13.6}{n^2}\text{eV}. \tag{6.5}$$

The unit of energy is the electron volt (eV) (see Chap. 5).

Bohr then made another bold assertion by requiring that electrons only change orbits by emitting or absorbing photons. Thus if an electron is to transition from the $n = 2$ orbit to the $n = 1$ orbit it must emit a photon of energy

$$E_2 - E_1 = -13.6\left(\frac{1}{2^2} - \frac{1}{1^2}\right)\text{eV} = 10.2\,\text{eV}. \tag{6.6}$$

Using the equation $E = hf$ shows that this corresponds to light with frequency $f = 2.47 \cdot 10^{15}$ Hz, which is somewhere in the ultraviolet portion of the spectrum. Notice that Bohr's idea is in agreement with the law of conservation of energy.

Bohr's model was immediately accepted because it explained *spectral lines*. Recall from Chap. 5 that the emission and absorption of light by hydrogen is described by Rydberg's formula (Eq. 5.6), which is precisely of the form given above. More remarkably, the factor in Bohr's energy formula, $me^4/(8h^2\epsilon_0^2)$, agrees precisely with the Rydberg constant in Rydberg's formula!

6.6 The Quantum Atom

Although Bohr's quantum model of the atom explained the Rydberg formula, it was not perfect. First, electrons still moved in circles and still obeyed Maxwell's equations, so they should still lose energy by emitting electromagnetic radiation. Second, the agreement with Rydberg's formula proved too good because it was soon realized that the experimental spectrum has small, but definite, corrections. Finally, Bohr's model did not generalize well to atoms with many electrons, where its agreement with experimental results was not good.

Although Bohr's model proved to be wrong, it introduced two crucial ideas: atomic energy is quantized and the interactions of photons with atoms requires electrons to change their quantum numbers. Most importantly, it extended the quantum concept from light to matter.

This idea was formalized in 1924 by Louis, the 7th Duc de Broglie (1892–1987).[8] De Broglie was convinced that if light was particle-like sometimes then electrons (and other particles) should be wavelike sometimes. He formalized this guess by taking the Planck-Einstein relationship, $E/c = h/\lambda$ and writing it for particles as

$$p = \frac{h}{\lambda}. \tag{6.7}$$

---

[8]Pronounced "de broy".

In this equation λ is the wavelength of a particle and p is its momentum (you might know the equation $p = mv$).

Physicist Walter Elsasser (1904–1991) suggested that it should be possible to test the idea by looking for diffraction when scattering electrons off of a crystalline solid. The experiment was soon completed by Clinton Davisson (1881–1958) and Lester Germer (1896–1971) at Bell Labs in New Jersey (Fig. 6.5) and dramatically confirmed de Broglie's idea.

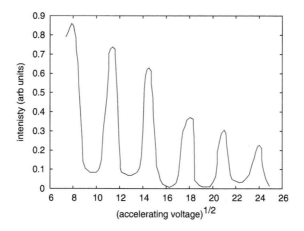

Figure 6.5: Electron scattering intensity from nickel.

Source: C.J. Davisson, *Are Electrons Waves?*, Franklin Institute Journal **205**, 597 (1928).

De Broglie's suggestion also prompted the world's leading physicists to find a reasonable description of atoms and their spectral lines. Perhaps it should not be surprising that the problem was solved twice within a year (1925–1926) by completely different methods. The first solution was due to a brilliant newly minted PhD, Werner Karl Heisenberg (1901–1976), who abstracted classical Newtonian mechanics in a profound – and to most people at the time, profoundly confusing – way. The kernel of Heisenberg's idea was to replace Newtonian position and velocity with infinite size matrices.[9] This idea came to be known as ***matrix mechanics***.

Austrian physicist Erwin Rudolf Josef Alexander Schrödinger (1887–1961) provided the second solution following more conventional methods. Schrödinger thought that if the electron were a wave it should satisfy a wave equation. But what equation? Eventually he realized that Newtonian and wave descriptions of matter could be combined if he mapped classical quantities like momentum and energy to derivatives (rates of change) of ***wavefunctions***. This led to his famous

---

[9]A ***matrix*** is a two-dimensional array of numbers.

Schrödinger equation and to the new field of **quantum mechanics**. Again, if you want to put it on a t-shirt, the equation reads[10]

$$i\hbar \frac{\partial}{\partial t}\psi(x,t) = -\hbar^2 \frac{\nabla^2}{2m}\psi(x,t) + V(x,t)\psi(x,t). \qquad (6.8)$$

Notice that both Heisenberg and Schrödinger were forced to replace Newtonian quantities (position, velocity, energy) with more general concepts (matrices or derivatives of wavefunctions). This idea was formalized by Bohr and named the **correspondence principle**. The correspondence principle is designed to describe matter as wave-like, but reduce to particle-like behavior when systems get large, just as particle-like photons reduce to wave-like electromagnetic radiation when their numbers get large.

A curious feature of quantum theory is that it is not possible to determine the exact position of an electron (think about specifying the position of an ocean wave). Because of this, one can no longer speak of electron orbits, like in the Rutherford or Bohr models, but of a probability distribution of an electron's location. The result is that the concept of orbits is replaced by that of **orbitals**, which you can think of as fuzzy orbits that specify the probability of finding an electron with a given set of quantum numbers.

Within the year, the predictions of Heisenberg's and Schrödinger's theories had been obtained for the hydrogen atom and were in complete agreement with Rydberg's formula. Of course this immediately raised the suspicion that the theories were in fact equivalent, which was soon established by several people. The early successes of the new quantum mechanics were followed by a flood of applications and ideas as a generation of physicists mapped out the new quantum world and found quantum explanations for a vast array of small scale phenomena. These included the structure of metals, the nature of electricity, molecules, chemical reactions, the structure of the nucleus, the workings of the sun, nuclear radioactivity, and the scattering of light.

6.7 Nuclear and Atomic Scales

There are substantial differences between an atom and the nucleus of that atom that set important **scales** for some of the issues we discuss in the following chapters. A scale is something that sets a reference size (length, energy, time, etc) for a given problem. For example, if we are concerned with sending a rocket to Mars the relevant scales are masses of planets and the distances between them. Phenomena on much smaller scales – say me jumping up and down – are negligible to this problem.

---

[10]The notation $\hbar$ is Planck's constant divided by 2π. V is the potential energy of the system. The quantity ψ is called the *quantum wavefunction* and represents the future behavior of the system in probabilistic terms.

Chemical properties of atoms are determined by electrons in outer orbitals. Thus the length scale of chemical processes is angstroms (10^{-10} m).[11] And, in case you were wondering, *this* scale is set by the mass of the electron.[12] De Broglie's relationship implies that an analogous energy scale is given by Planck's constant divided by a length scale. For the atom this works out to 2 keV. When detailed calculations are made the actual energy scale is about 100 times smaller (see Eq. 6.5), coming in at tens of electron volts.

In contrast, the nucleus is comprised of particles that are around 2000 times heavier than the electron and occupy a space that is 100,000 times smaller than the atom (this is the ratio for the radius). Thus the natural nuclear length scale is femtometers (1 fm = 10^{-15} m) and the natural energy scale is tens of millions of electron volts (MeV). The enormous ratio of energy scales between nuclear and chemical phenomena is an indication that vast energies lie in the nucleus, if only they can be extracted.

6.8   The Interaction of Light and Matter

Photons are the quantum carriers of electromagnetic waves and the electromagnetic force. Atoms and molecules and people are built from electrons, protons, and neutrons held together by the electromagnetic force,[13] and we interact with the rest of the universe by this same force. Touch is sensitive to pressure created by repulsive electromagnetic forces, hearing detects sound waves created by similar forces between air molecules, our eyes are devices that detect photons scattered off of distance objects, and the resulting signals are interpreted by a brain that is powered by chemical electromagnetic reactions. We explore some of these interactions in greater detail in this section.

The quantum nature of matter means that atoms and molecules can only absorb light at certain discrete frequencies. This leaves the atom in an ***excited state***, where an electron has been promoted to a higher orbital. This electron will then spill off energy in the form of photons and drop back down to the lowest orbital. Energy must be conserved during this process, which can be written as a formula:

$$[hf + E_{\text{electrons}} + E_{\substack{\text{atomic} \\ \text{motion}}}]_{\text{before}} = [hf + E_{\text{electrons}} + E_{\substack{\text{atomic} \\ \text{motion}}}]_{\text{after}}. \qquad (6.9)$$

Notice that the *kinetic energy* – the energy of motion – of the atom has been included in the formula. Since the electrons typically start and end in the lowest orbitals, one sees that the change in the photon energy is balanced by a change in the speed of the atom. The net result is that the light has been ***scattered*** with a possible change in energy, direction, and polarization, while the atom may have recoiled, either picking up or losing kinetic energy.

---

[11] Atoms range in size from 0.3–3.0 Å.

[12] What sets the electron mass scale is unknown.

[13] Except within the nucleus, which is the topic of Chap. 8.

6.8.1   Absorption of Light

As light moves through a transparent medium like water it continually interacts with H_2O molecules and is re-emitted at a lower frequency while imparting kinetic energy to the H_2O molecule. This means that the molecules are moving faster, which means that the water is heating up. At sufficient depth the light will have faded away, leaving only warmer water.

The H_2O molecules absorb and re-emit at their characteristic energies and this is reflected in how the light fades with distance into the water. A measurement of this absorption is shown in Fig. 5.8 and reveals a startling fact. As you can see from the figure, water readily absorbs light between 10^6 and 10^{18} Hz, *except* for a small band of frequency which coincides exactly with the visible spectrum. In view of the Theory of Evolution, this is a remarkable demonstration of the watery origins of our species.

6.8.2   Why the Sky Is Blue and Milk Is White

In general the scattering of light from objects can be complicated. However, when the scattering object is much smaller than the wavelength of the light, the situation simplifies and a formula can be obtained. This was first figured out by John William Strutt, 3rd Baron Rayleigh (1842–1919), who was Thomson's predecessor as head of the Cavendish Lab. Rayleigh's formula shows that the intensity of the scattered light scales as $1/\lambda^4$; in other words, light with smaller wavelengths scatters more readily than light with longer wavelengths.

This simple result explains why the sky is blue. First, on clear days light scatters off of molecules in the air. These are roughly 1/10 of a nanometer in size, hence they are much smaller that visible light (400–700 nm), and Rayleigh's formula applies. Imagine you are looking up on a sunny day. Light from the sun is streaming into the atmosphere and scattering off of air molecules. Light is scattered into all directions, and some of it is directed toward your eyes. But blue is scattered more than red, so the sky appears blue.

Alternatively, if you look at the horizon at sunset, the light must travel directly to you. In this case blue light is scattered away and it is reddish light that reaches your eyes. So sunsets are red for the same reason the sky is blue!

When objects are much bigger than the wavelength of light, geometric optics describes the scattering process. And when objects are about the same size as the light wavelength, the situation gets complicated, but generally light of all wavelengths is scattered equally. If the incoming light is white, the scattered light will be white. Thus the sky looks much paler when there are a lot of water droplets in the air. Similarly, milk looks white because it contains droplets of fat that are in the 400–700 nm range in size. Maybe you have noticed that skim milk looks bluish. Can you find an explanation?

Ex. Smoke curling from the end of a cigarette has a bluish tinge, so it must contain many very small particles. On the other hand, exhaled smoke is white. We conclude that small particles are being trapped in lung tissue while large particles (water vapor) are being added. This is a sure sign of danger! The small particles are absorbed into cells lining the lungs and disrupt cell function, eventually causing cancer.

6.8.3   Polarization

Light can be polarized by scattering, especially from surfaces like glass or water. One can construct a simple classical (i.e., not quantum) model to understand this effect. Imagine that an atom consists of a heavy nucleus and a small charged ball attached to it by a spring. When an incoming electric field (i.e., an electromagnetic wave) impinges the atom it sets the charge in motion along the direction of the field, and the moving charge creates another electric field which forms part of a *scattered wave*.[14]

It turns out that reflected light is polarized because the scattered electric field is zero along the direction of charge oscillation. To see how the polarization occurs we will imagine unpolarized incident light that is broken down into two components, these are referred to as $\vec{E}_{||}$ and $\vec{E}_{\perp}$ in Fig. 6.6. The grey arrow in the left figure shows the direction in which the electric field oscillates. This induces oscillation in the scattering atom that is shown by a black double-headed arrow. As just mentioned, the resulting electric field does not radiate in the direction of the observer and hence is not seen.

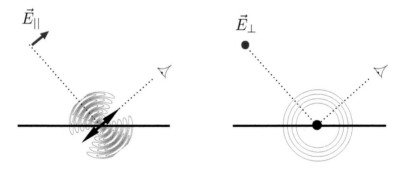

Figure 6.6: The polarization of light by reflection.

The situation is quite different for the other component of the incident electric field. The grey circle in the right figure indicates that the electric field $\vec{E}_{\perp}$ is oscillating on a line that goes into (and out of) the page. This induces an oscillation

---

[14]This, by the way, is how the world scatters light into your eyes and allows you to see.

in an atom that is also into (and out of) the page, which is denoted by the black circle. Again, no radiation will be emitted on a line that is at right angles to the page, but radiation *is* emitted along the line to the observer. Thus the observer only sees radiation that is polarized in a plane along the line of sight and at right angles to the page.

I am sure that you are familiar with glare from the surface of water, ice, or snow. This is due to reflected light that, as we just learned, is horizontally polarized. Thus wearing vertically polarized sunglasses eliminates reflected sunlight, which makes it easier to see other things.

> Look at an LCD screen while wearing polarized sunglasses. Slowly rotate the glasses (or screen) and observe carefully. What do you conclude?

6.8.4 Nonionizing Radiation and Human Health

You are probably aware that it is UV radiation from the sun that causes tanning and skin cancer. Yet if you spend all day under an artificial light you do not tan and you do not get skin cancer. This is because artificial lights do not emit very much radiation above the visible band of the spectrum. We conclude that electromagnetic radiation with energy above about 10 eV is dangerous while radiation with lower energy is safe.

> Ex. X-ray radiation lies above UV and can cause cancer. Gamma rays are higher energy and even more dangerous.

Can we understand this? Take another look at Eq. 6.5 and Eq. 5.6. When the integer gets very large, the electron is in an extremely excited state, and its average distance from the nucleus becomes very large. As n goes to infinity, the electron is dissociated from the nucleus and the atom is said to be *ionized*. Rydberg's formula implies that this can happen when a photon of energy 13.6 eV or greater strikes the atom. All electromagnetic radiation with energy greater 13.6 eV is called *ionizing radiation* while electromagnetic radiation with energy less than 13.6 eV is called *nonionizing*. (These numbers are for hydrogen; other atoms and molecules have similar ionization thresholds.)

The answer to our question is now apparent. UV radiation is dangerous because it is ionizing. In other words, it has sufficient energy to disrupt biomolecules in the skin by stripping them of electrons. If these molecules happen to be DNA, the damage can interfere with DNA function and cause cancer. Alternatively, if the light has energy in the visible range or lower, it does not cause damage.

Importantly, the intensity of nonionizing radiation does not matter when it comes to cancer. We know this because of the photoelectric effect. Recall that electrons could not be dislodged from a metal below a threshold energy, *regardless of the intensity of the light.* An analogy would be trying to throw a rock across a river. If the river is too broad, it does not matter how many times you try, you will not be able to get one across.

> Ex. Cell phones emit radiation in the 800–2500 MHz range. This is around one million times too weak to disrupt biomolecules.

Nonionizing radiation does not cause cancer.

REVIEW

Important terminology:

the Big Bang   [pg. 128]

black body, blackbody radiation   [pg. 126]

electron, proton, neutron, nucleus   [pg. 125]

ionizing radiation   [pg. 137]

orbital   [pg. 133]

quantum, quantization, quantum number   [pg. 130]

scale   [pg. 133]

Stefan-Boltzmann law, $j = \sigma T^4$ [pg. 128]

Important concepts:

Wave-particle duality.

The correspondence principle.

The plum pudding model.

The Bohr atomic model.

The Rutherford atomic model.

The quantum atom.

Blackbody radiation universality.

The photoelectric effect.

Accelerating charges emit electromagnetic radiation.

Electrons change atomic orbitals by emitting or absorbing photons.

Lower wavelength light scatters more strongly off of small particles.

Polarization by reflection.

FURTHER READING

S. Schweber, *QED and the Men Who Made It*, Princeton University Press, 1994.

EXERCISES

1. Units.

   Using Eq. 6.1, verify that the units of Boltzmann's constant are energy/temperature. Determine the units of U.

2. Magnetic Fields.

   Look again at Fig. 5.3. You would think that the magnet should create a field that is smooth, so why do the filings tend to line up with spaces between the lines?

3. Rydberg constant,

   Confirm that Bohr's formula agrees with the Rydberg formula Eq. 5.6.

4. Hydrogen Transition.

   What is the frequency of light associated with a transition from the 5th to 3rd orbital in hydrogen?

5. Polarized Diving Googles.

   Discuss how light would be polarized under water. Would it be useful to manufacture diving masks with polaroid lenses?

6. Electrons.

   Can electrons

   (a) diffract

   (b) reflect

   (c) refract?

7. Red and blue scattering.

   If red light scatters a certain amount, how much more does blue light scatter?

8. Scattering.

   An atom of mass $1200\,\text{MeV}/c^2$ is at rest when it absorbs a photon of frequency $1.4\,\text{THz}$. It then emits a photon of frequency $0.8\,\text{THz}$. Assuming that it starts and ends in the ground state ($n = 1$), how fast must the atom be moving after emitting the photon?

9. Two Polarizers

   Polarize a light source with a polaroid film or sunglasses. Now look at that light through another polarizer. What do you expect to see if the polarizers are

   (a) aligned

   (b) at 90 degree to each other

   (c) at 45 degree to each other?

10. Three Polarizers

    Polarize a light source, look at this with a second polarizer at right angles to the first. You should see nothing. Now insert a third polarizer between the two at 45 degree. What do you expect to see? Read up on how LCD screens work and relate what you find to this question.

11. Dangerous Radiation.

    Look up the frequencies at which the following devices work and decide if they are dangerous.

    (a) microwave ovens

    (b) wifi

    (c) power lines.

12. Cosmic Background.

    The relic light from the early universe (Fig. 6.3) is called the "cosmic microwave background". If its current temperature is 2.7 K, what do you think its temperature was when the universe was 1000 times more compact? Why? Why is it called a "microwave" background?

13. De Broglie Equation.

    In Sect. 6.6 the formula $E/c = h/\lambda$ was used. Where did this come from?

Climate

"Climate is what we expect, weather is what we get."

In Chap. 4 we learned that we are rich in good part because energy is cheap and abundant. But the party cannot go one forever (Chap. 9) and there are substantial costs that we are currently ignoring. Chief among these is the fantasy that pumping billions of tonnes of carbon into the air does not affect us or the climate. In this chapter we will look at the real costs associated with burning coal and oil. We then examine the effects this has on energy flow through the environment, and hence on the climate. Finally, climate models and their predictions are examined.

7.1 Fossil Fuels

7.1.1 Burning Coal and Oil

It is estimated that 9.5 Gt of carbon was emitted by fossil fuel combustion and cement production in 2011. Land use (such as deforestation) contributed an additional 0.9 Gt.[1]

The total emissions since the industrial revolution began total around 365 Gt (Fig. 7.1). Deforestation over the same span has contributed an additional 180 GtC.

Tracking this carbon is vital to understanding its effect on the environment (and us). American geochemist Charles Keeling (1928–2005) made an important contribution to this by developing a method to measure atmospheric CO_2 concentration. A long series of subsequent measurements at the Mauna Loa Observatory in Hawaii (Fig. 7.2) have been credited with raising awareness of the issue. In particular an obvious growth of about 2 ppm/yr in CO_2 concentration is visible in the data.

By 1960 Keeling had collected enough data to surmise that the increase was due to coal combustion.

---

[1]One often sees figures for carbon emissions or CO_2 emissions. Since carbon has an atomic weight of 12 and oxygen has atomic weight 16, 1 Gt carbon is equivalent to 44/12 Gt CO_2.

© Springer International Publishing Switzerland 2016   143
E.S. Swanson, *Science and Society*,
DOI 10.1007/978-3-319-21987-5_7

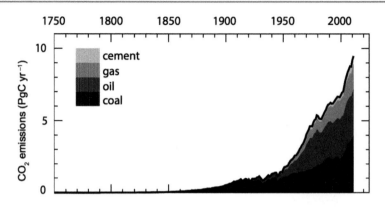

Figure 7.1: Global CO_2 emissions.

Reproduced with permission, IPCC AR5. Ciais, P., C. Sabine, G. Bala, L. Bopp, V. Brovkin, J. Canadell, A. Chhabra, R. DeFries, J. Galloway, M. Heimann, C. Jones, C. Le Quéré, R.B. Myneni, S. Piao and P. Thornton, 2013: Carbon and Other Biogeochemical Cycles. In: Climate Change 2013: The Physical Science Basis. Contribution of Working Group I to the Fifth Assessment Report of the Intergovernmental Panel on Climate Change [Stocker, T.F., D. Qin, G.-K. Plattner, M. Tignor, S.K. Allen, J. Boschung, A. Nauels, Y. Xia, V. Bex and P.M. Midgley (eds.)]. Cambridge University Press, Cambridge, United Kingdom and New York, NY, USA, pp. 465–570, doi:10.1017/CBO9781107415324.015.

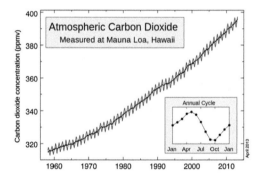

Figure 7.2: Atmospheric carbon dioxide (ppm by particle).

Ex. Let's check this. We currently emit 9.5 GtC/yr which is 10^{16} g C/yr or $10^{16}/12 \cdot N_A$ CO_2 molecules per year. It is estimated that the total mass of the atmosphere is $5 \cdot 10^{21}$ g, which is 78% N_2 and 21% O_2, giving a total of 10^{44} atmospheric particles. The increase rate is thus about 6 ppm. About ⅓ of emissions accumulate in the atmosphere, so this works out just right.

Your personal contribution to this is about 5500 kgC per year (see Ex. 1), which amounts to 30 pounds of carbon per day. Much of this total arises because we like to live far from where we work and we don't like sharing transit. More specifically, burning gasoline releases (see Ex. 2)

$$8900 \text{ g } CO_2/\text{gal.} \tag{7.1}$$

Thus a car that gets 30 miles per gallon emits 176 g CO_2/km. The average US car is slightly less efficient than this:

> the average car emits ¾ pounds of CO_2 per mile.

Our trips to the grocery store carry a cumulative impact on the environment that we all should be acutely aware of.

Oceans are another possible sink of anthropogenic carbon.[2] Once in the water some of it reacts to form carbonic acid, thus it is possible to follow carbon uptake by measuring oceanic pH levels. Recent results indicate that the pH of surface water has decreased by about 0.1 (this corresponds to 26 % increased acidity) since the beginning of the industrial era.

Measurements such as these permit estimates of the disposition of the carbon produced by our energy habit. The results are shown in Table 7.1.

Table 7.1: Disposition of Historical Anthropogenic Carbon (including from deforestation).

Sink	Amount (GtC)
Atmosphere	240
Oceans	155
Land	150
Total	545

7.1.2 The Carbon Cycle

Determining the impact of anthropogenic carbon requires an in-depth understanding of the *carbon cycle*, which is the flow of carbon in the environment.[3] This is split into the *slow carbon cycle*, that deals with long scale absorption and emission of carbon in rocks, and the *fast carbon cycle*.

A simple representation of the fast carbon cycle is shown in Fig. 7.3. Black arrows indicate flows of carbon (called *fluxes*) with attached numbers that give the

---

[2] Anthropogenic means "made by man".
[3] Similarly, one speaks of the water cycle, the nitrogen cycle, and so on.

flux in GtC/yr. Lighter numbers are human contributions to the flow of carbon, while the numbers in cubes represent stored carbon in Gt. We see that photosynthesis dominates carbon fluxes on land. Plants take up about 123 GtC/yr and convert this into plant mass and soil carbon[4], while respiring about 60 GtC/yr back to the atmosphere. Decomposition and microbes also put about 60 GtC/yr into the atmosphere, bringing the cycle (nearly) into equilibrium.

The primary activity over the oceans is direct absorption and emission of gaseous carbon-dioxide by water. Photosynthesis by plankton and algae is offset by respiration. Finally, there is a slow leakage of carbon into the ocean depths at about 0.2 GtC/yr which ultimately goes into the formation of limestone.

It is important to recognize that carbon reservoirs are not *static* on any time scales. For example, the Keeling curve (Fig. 7.2) shows seasonal variation in atmospheric carbon. Similarly, upswelling changes ocean carbon content over time spans of 100k years and volcanic activity changes rock carbon content over scales of 100 million years. Nevertheless, carbon reservoirs tend to be stable because they are subject to *negative feedback*. Feedback is a process where a small change in a system causes subsequent changes that either reinforce the initial change (*positive feedback*) or inhibit the initial change (negative feedback).

> Ex. Positive feedback: increased temperatures in the arctic melt snow, exposing dark soil, which absorbs heat readily, which increases temperatures.

> Ex. Negative feedback: a ball in a valley is given a small push, rolling it up the hill, where gravity causes it to roll back down, and friction causes it to slow down.

7.2 Greenhouse Gases

You have probably heard that greenhouse gases are Earth's quilt. Although this is the right sentiment, the full story is more complicated.

7.2.1 Black-Body Earth

As a simple starting point, let's model the Earth as a featureless ball with no atmosphere (I mean *air*, not *night life*). The sun illuminates the Earth with a fixed intensity of light (we can measure this in W/m^2). This energy heats the Earth, and because there is no atmosphere this heat cannot escape to space by convection or conduction. This leaves infrared radiation as the only mechanism for the Earth to lose energy. Hot Earth molecules bounce around more, which means they are accelerating, which means they emit electromagnetic radiation (see Sect. 6.5).

---

[4]Recall the estimate of 100 Gt/yr we made for this in Sect. 4.4.2.

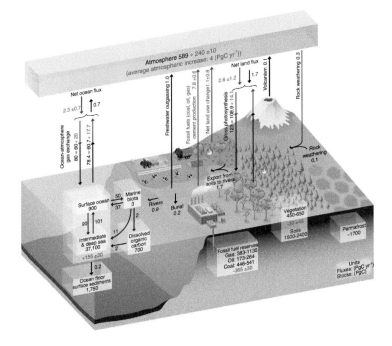

Figure 7.3: Simplified fast carbon cycle.

Reproduced with permission, IPCC AR5, see Fig. 7.1.

Perhaps it is not surprising that this radiation is in the infrared thermal part of the spectrum (100 μm wavelength, 0.01 eV energy, THz frequency).

We will assume that the sun emits radiation as a black-body (Fig. 7.4 shows this is a good assumption), thus its energy is emitted according to Planck's law and its total emitted power per unit area is given by the Stefan-Boltzmann law $j = \sigma T^4$ (see Sect. 6.3). Once the Earth's temperature equilibrates, its temperature must be such that its outgoing power in thermal radiation matches the incoming, $P_{in} = P_{out}$. We will assume that the Earth also emits as a black-body so that $P_{out} = 4\pi R_\oplus^2 \sigma T_\oplus^4$ (where $\oplus$ refers to the Earth). The part before σ is the surface area of the Earth.

The incoming power is the sun's power times the fraction of the sun's radiation that intercepts the Earth, $\pi R_\oplus^2/(4\pi R^2)$, where R is the average distance between the sun and Earth. Now some of this electromagnetic radiation is reflected back to space. The ratio of outgoing radiation to incoming is called the ***albedo***. Albedo ranges from 0 for black objects to 1 for perfectly reflective white surfaces.[5]

---

[5]Typical albedos are: bare soil (0.17), grass (0.25), sand (0.40), snow (0.85).

Putting it all together gives ($\odot$ stands for the sun; α is the value of the albedo)

$$\sigma T_\odot^4 4\pi R_\odot^2 (1 - \alpha) \cdot \frac{\pi R_\oplus^2}{4\pi R^2} = \sigma T_\oplus^4 4\pi R_\oplus^2. \qquad (7.2)$$

Using $T_\odot = 5778\,\text{K}$, $\alpha = 0.30$, $R_\odot = 6.96 \cdot 10^8\,\text{m}$, and $R = 1.496 \cdot 10^{11}\,\text{m}$ gives $T_\oplus = 254\,\text{K} = -19\text{C}$.

Notice that the result is independent of the size, mass, or composition (assuming the same albedo) of the Earth. Thus there is no way out of this startling result: the Earth would be very cold indeed if it weren't for its atmosphere. There would be no open oceans and it is very unlikely that life would have evolved at all. Our atmosphere is indeed a comfy quilt!

7.2.2   Greenhouse Gases and Global Temperature

The atmosphere saves the day for humanity because it is full of trace gases that absorb and re-emit thermal radiation. Some of this radiation is reflected back to Earth, where it warms the planet, and ultimately attempts to make its escape again. The net result is that the surface of the Earth averages a comfortable 15 C.

Direct measurement of the intensity of light, shown in Fig. 7.4 (this is referred to as **insolation** and is measured in watts per unit area per unit frequency), confirms these general features. The figure shows the spectrum of light as emitted

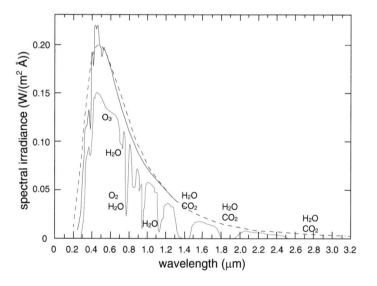

Figure 7.4: Sunlight spectra at sea level and at source.

Source: *Handbook of Geophysics and Space Environments*, S. Valley (ed.), 1965, McGraw-Hill. Image by M. Siegert-Wilkinson.

by the sun as a solid line. Notice that it is quite similar to the theoretical curve (dashed line) for a black body emitting at 5900 K. Notice also that the peak of the sun's power is right in the visible range of the spectrum (0.4–0.7 μm). It is thus quite convenient that water happens to be transparent to this portion of the spectrum as well! (see Sect. 5.6)

The lower green line is the measured intensity of light at sea level (i.e., after it has passed through the atmosphere). One sees a general decrease of the intensity, indicating that the atmosphere is absorbing or reflecting light, and an intricate pattern of absorption bands where the passage of light is severely attenuated. The molecules responsible for many of these bands are shown in the figure. These molecules are the **greenhouse gases**. The figure makes it clear that water vapor is an important greenhouse gas!

We have learned that the emission and absorption of light by atoms depends on their electronic structure (i.e., atomic electron orbitals). Energies associated with electronic transitions are typically in the electron volt range, which corresponds to visible or UV light, and is too high energy for scattering light in the infrared part of the spectrum.

But greenhouse gases are not atoms, they are molecules, and molecules permit more complicated motions than atoms. Consider a molecule of carbon dioxide. This is a linear chain, O=C=O, and this chain can bend, or the lengths of the bonds can oscillate, or the whole molecule can rotate. These motions are illustrated in Fig. 7.5. Chemists distinguish three types of molecular quantized motion: electronic, vibrational, and rotational. Vibrational electromagnetic transitions are typically 1–30 mm (visible and IR light), while rotational are 100–10,000 mm (thermal to microwave).

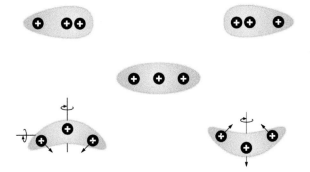

Figure 7.5: Modes of oscillation of CO_2.

Image by E. Cochran.

Thus it is the rotational modes of greenhouse gas molecules that permit transitions in the infrared part of the spectrum; these in turn keep the planet from looking like a giant snowball.

Ex. It is possible you have personal experience with the importance
of greenhouse gases in reflecting thermal radiation. Desert air is very
dry and nearly transparent to infrared radiation. Once the sun goes
down the heat of the day simply radiates into space, and deserts get
very cold at night.

Notice that vibrational oscillation and bending in CO_2 both displace the positive charge of the nucleus (and hence, since the molecule is neutral, the negative charge of the electrons). In this way the charge is made "visible" to the electromagnetic field and scattering can take place. Alternatively, symmetric diatomic molecules like H_2, N_2, and O_2 can't reveal their hidden electric charge because vibrational oscillation just changes the distance between the nuclei and bending is impossible. Thus these molecules, although common in the atmosphere, have little to do with greenhouse heating.[6]

As a result the list of important greenhouse gases is rather short, and in fact, carbon dioxide plays the dominant role. This is shown in Table 7.2, where we see that the amount of carbon dioxide and its *radiative forcing* dominate.

Table 7.2: Greenhouse gases and their effects.

Gas	Historical concentration	Current concentration	Increase in radiative forcing
Carbon dioxide (CO_2)	280 ppm	401 ppm	1.88 (W/m^2)
Methane (CH_4)	0.722	1.80	0.49
Nitrous oxide (N_2O)	0.270	0.325	0.17
Ozone (O_3)	0.237	0.337	0.40
Halocarbons			0.36

Source: IPCC AR5.

7.3 Energy Flow in the Environment

Radiative forcing, measured in power per unit area, is an indication of how much influence a greenhouse gas (or other factor) has on the climate. The idea is that the temperature of the Earth is driven by the rate at which it receives solar energy (this is the left hand side of Eq. 7.2). If we divide that by the surface area of Earth we get an average incoming energy of (we neglect the albedo for now):

$$340\frac{W}{m^2} \quad \text{[incoming solar power density]}. \tag{7.3}$$

---

[6]They *can* become important if their electronic structure is disturbed by collisions, which can happen at sufficiently high temperatures and densities.

The total additional radiative forcing due to the heating effect of greenhouse gases is $3.3\,\text{W/m}^2$. This is only 1 % of the incoming power density – but it is an important 1 %, as we shall see!

Figure 7.6 shows the main flow of energy through the environment. We see that about 30 % of the incoming energy is reflected back to space by clouds and the ground. Thus the Earth's albedo is about 0.3, in agreement with the number we used in Eq. 7.2. The other 70 % goes into heating the atmosphere and the surface. This heat ends up as infrared radiation that is absorbed and emitted repeatedly in the atmosphere, until it ultimately escapes back to space. In this way the incoming energy is balanced by the outgoing energy and the Earth maintains a constant temperature. As we have discussed, this temperature is higher than it would be without greenhouse gases slowing the outward flow of energy.

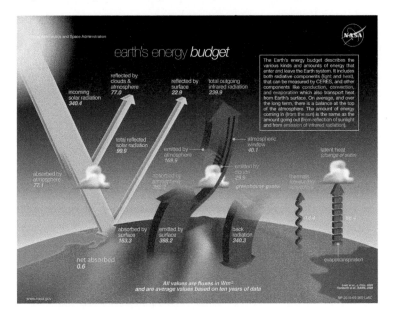

Figure 7.6: Flow of energy in the environment.

What happens when carbon dioxide is added to the atmosphere? At a simple level, the added CO_2 makes it more difficult for thermal radiation to escape and the temperature of the Earth rises, which creates more thermal radiation, that finally rebalances the energy flow.

Importantly, thermal radiation tends to get absorbed and re-emitted by layers of the atmosphere, making its way to the (thin and cold) top of the atmosphere, where it finally escapes. Adding more CO_2 increases the effective height of the "escape layer", which means that additional CO_2 always has an effect on the energy budget of the planet.

Additional carbon dioxide always warms Earth.

This is only possible because carbon dioxide is a *well mixed* gas, which means
it quickly disperses throughout the atmosphere. Gases that are not well mixed
tend to react with rocks or other parts of the environment and leach out of the
atmosphere.

The extra carbon dioxide has a number of effects on the Earth that we will
explore in Sect. 7.5. The most famous of these is global warming. Figure 7.7
shows direct and indirect reconstructions of the average global temperature from
1000 AD. Modern measurements are based on readings at observatories around
the world (from 1900 on), while older ones are based on *proxies* such as tree ring
data, and ice core samples. The gray band indicates the error in the data. A rapid
rise in temperature after 1900 is evident and gives this graph its name, "the hockey
stick".

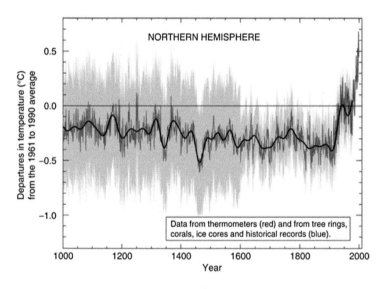

Figure 7.7: Global temperature anomaly.

Reproduced with permission, IPCC AR3. Climate Change 2001: The Scientific Basis. Contribution of
Working Group I to the Third Assessment Report of the Intergovernmental Panel on Climate Change,
Technical Summary, Figure 5. Cambridge University Press.

Sometimes people say that similar temperature excursions have happened in
the past. The point, of course, is that for the first time in history *we* are causing
a temperature anomaly. It is up to us to decide what, if anything, we should do
about it.

7.4 Climate Modelling

How do we know that the global temperature rise is due to carbon dioxide emissions? Recall that the radiative forcing due to anthropogenic CO_2 is about 1 % of the total. Since natural greenhouse gases heat the Earth about 35 degrees C, it is not unreasonable to expect a 0.35 C additional rise in temperature. The hockey stick graph indicates a rise of about 0.7 C so this is not too far off!

What we are missing in our simple estimate is *feedback mechanisms*. The most important of these are a negative feedback due to increasing thermal radiation with temperature (according to the Stefan-Boltzmann law), and a positive feedback due to increasing evaporation of water, which traps heat. Properly accounting for all energy flow and feedback is the business of *climate models*.

At this stage it is important to emphasize that climate and weather are different things.

> Climate is not weather.

Weather is what we experience on a daily basis. It is difficult to predict because local weather patterns depend on many variables, most of which are poorly known. Weather is also chaotic (see Chap. 1). *Climate* refers to long term systematics in the Earth's weather. These are easier to predict because randomness is averaged out when long time scales (or large areas) are considered.

Climate scientists use well-known laws of physics to model planetary climates (not just the Earth's!). This is not an easy task! Scientists must solve equations of heat flow in the atmosphere and oceans. These are turbulent systems that require detailed numerical computations that tax the fastest of our computers. Many biological, chemical, and geological processes are also accounted for. Amongst these are cloud formation, jet streams, ocean streams, plant photosynthesis and respiration, forest use, farm land use, rainfall, methane production by animals, rock weathering, volcanic activity, crop fertilization, particulate pollution production, ice cap melting, permafrost thawing, and hundreds of more processes.

The fundamental difficulty in climate modelling is the vast range of scales, both temporal and spatial, that are present. Jet streams and hurricanes are continent-sized, while the scattering of light and creation of ozone occur at the atomic scale. This represents a range of 16 orders of magnitude! In practice, climate modelers would like to be able to simulate clouds while building up a representation of the entire planet. Thus it is desirable to divide the atmosphere into about $4 \cdot 10^{12}$ cubes of size $(100\,\mathrm{m})^3$. Unfortunately this is impossible with modern computers – current climate models divide the atmosphere into about $180 \times 360 \times 40$ cubes.[7]

---

[7]Forty slices are taken in the vertical direction, while the horizontal direction represents squares of size $2° \times 1°$.

Every 6 years the United Nations charges the Intergovernmental Panel on Climate Change (IPCC) with reviewing the current climate data and models and issuing a consensus report. The 2014 report, called the Fifth Assessment Review (AR5), was written by more than 800 authors with contributions from more than 2500 scientific reviewers and runs into thousands of pages. It is available online so that you can judge for yourself whether these scientists were thorough in their analysis.

Of course, it is energy from the sun that drives all of the considered processes.[8] Solar insolation can vary due to internal solar dynamics (like the sunspot cycle) and due to perturbations in the Earth's orbit. The most famous of these are called *Milanković cycles*, which are periodic variations in the shape of the Earth's orbit.[9] While a prisoner of war during World War I, Serbian astronomer Milutin Milanković (1879–1958) proposed that the resulting variation in solar insolation drove historical climate change such as the ice ages. This idea is now accepted, in part because modern ice core measurements reveal a variation of solar insolation over a period of about 100,000 years. These variations amount to $\pm 10\%$, with a resulting temperature fluctuation of about ± 2 C.

Once a climate model is made, it is tested and calibrated against historical data and then can be used to make predictions. We can get an idea of how accurate they are by looking at past predictions, such as that shown in Fig. 7.8. The figure shows the predicted rise in temperature under two scenarios for energy growth (red and blue curves). These predictions were made in 1981 based on only 30 years of past data. Actual temperatures up to 2011 are plotted as the wiggly line. We see that the fast growth scenario does a good job of predicting the subsequent temperature rise, even though climate models and computers were quite primitive 33 years ago.[10]

Particulate matter in the atmosphere, called *aerosols*, reflects sunlight and therefore lowers insolation, affecting the climate. The eruption of Mount Pinatubo in 1991 threw massive amounts of aerosols into the atmosphere and therefore provided an opportunity to test the ability of climate models to predict the resulting climate response. The models accurately predicted a drop in global temperature of about 0.5 C (see Fig. 7.9) and the subsequent rebound as the aerosols settled out of the atmosphere. Other quantities such as water vapor levels also agreed with model predictions, providing a dramatic verification of their accuracy.

Finally, we return to the question that opened this section: climate models confirm our simple estimates and show that the primary cause of recent global temperature rise is anthropogenic carbon dioxide emissions.

---

[8]Except plate tectonics and volcanism.

[9]To be precise, in the eccentricity, axial tilt, and precession.

[10]The paper does not specify the computational resources used, but the CDC7600 was a typical supercomputer of the day. This had 1/1000 the memory and speed of a modern smartphone.

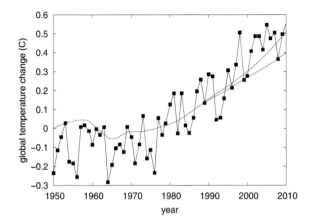

Figure 7.8: Global temperature rise as predicted in 1981. *Black points*: measured global temperature; *red* and *blue curves*: predicted temperature under different growth scenarios.

Source: J. Hansen, *et al. Climate Impact of Increasing Atmospheric Carbon Dioxide*, Science, 213, 957 (1981).

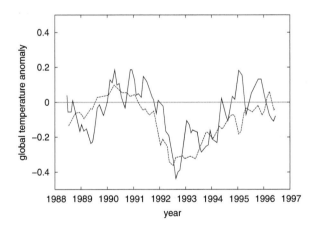

Figure 7.9: Observed and computed global temperature change during the Pinatubo eruption. The *solid line* is station data; the *dashed line* is the average model prediction.

Source: Data from J. Hansen *et al.*, *Climate Simulations for 1880–2003 with GISS model E*, Climate Dynamics **27**, 661 (2007).

7.5   Implications of Climate Change

Climate models are underpinned by physical laws and constrained by thousands of observations. They are certainly correct in capturing the main features of climate. Where they can be inaccurate is in details, like the rate at which clouds are formed or at which the polar ice caps break apart. The scientists are aware of the shortcomings in their models and incorporate corresponding errors in their predictions. Of course, future climate will depend on what we do. Since people are the least predictable part of climate models, climatologists present their predictions in terms of scenarios. The IPCC considered four scenarios, called Representative Concentration Pathways (RCPs): RCP2.6, RCP4.5, RCP6.0, and RCP8.5. The numbers represent radiative forcings in the year 2100. RCP2.6 assumes a tree-huggers paradise in which carbon emissions peak before 2020 and then decline substantially. Added carbon in this scenario amounts to 270 GtC. At the other end of the scale, RCP8.5 is a "drill baby drill" business as usual scenario. In this case an additional 1685 GtC carbon are assumed to have been burned by 2100. RCP4.5 and RCP6.0 are intermediate scenarios with corresponding carbon emissions of 780 and 1060 GtC (by 2100).

First let us consider the Milanković cycle. Data indicates that the insolation will be stable for the next 100,000 years, so there is little likelihood that we will slip into an ice age. The temperature cycle appears to be near a peak, and one can expect an orbitally forced drop in temperature of about 1 C over the next 25,000 years. We will see that this will be swamped by man-made temperature rise on much shorter scales.

7.5.1   Temperature Change

Temperature change is the most infamous of the effects of uncontrolled carbon emissions. Figure 7.10 shows the range of predictions out to 2100 for the RCP2.6 and RCP8.5 scenarios. The shaded areas represent model uncertainty. The prediction for the tree hugging scenario is a modest increase of 1 degree Celsius, which is stable after around 2040. Alternatively, the business-as-usual scenario predicts a global temperature rise of about 4 C (7.2 F) by 2100, with no indication of stabilization after that.

It is natural to think that a 4 C rise in temperature is not much of a concern. After all, we experience temperature shifts of 100 F or more between winter and summer in much of North America. But the 4 C rise is an average over the entire globe and our intuition can be misleading. A more familiar measure of the meaning of this warming is given in Fig. 7.11, which shows the predicted number of days per year one can expect temperatures over 95 F (median prediction for scenario RCP8.5). Notice that by 2100 some southern portions of the country will experience 200–250 days of hot temperatures per year!

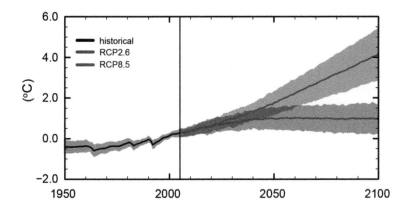

Figure 7.10: Past and predicted global average surface temperature changes.

Reproduced with permission, IPCC AR5. IPCC, 2013: Summary for Policymakers. In: Climate Change 2013: The Physical Science Basis. Contribution of Working Group I to the Fifth Assessment Report of the Intergovernmental Panel on Climate Change [Stocker, T.F., D. Qin, G.-K. Plattner, M. Tignor, S.K. Allen, J. Boschung, A. Nauels, Y. Xia, V. Bex and P.M. Midgley (eds.)]. Cambridge University Press, Cambridge, United Kingdom and New York, NY, USA, pp. 1–30, doi:10.1017/CBO9781107415324.004.

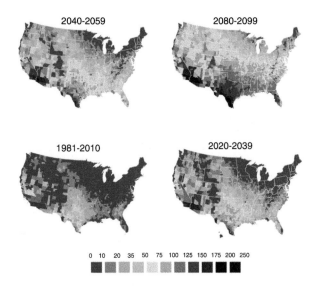

Figure 7.11: Predicted number of days over 95 degrees F.

Based on T. Houser, R. Kopp *et al.*, *American Climate Prospectus: Economic Risks in the United States.* New York: Rhodium Group LLC, 2014. Reproduced with permission.

7.5.2   Changes in Precipitation

A rising global temperature is not the only outcome of climate change. A warmer
Earth means more water vapor in the atmosphere, which implies more precipitation. The IPCC predicts increased rates of precipitation of 1–3 % per degree C in
all scenarios other than RCP2.6. Thus one expects about 8 % more global rainfall
and snow by 2100 (RCP8.5).

This rainfall is not uniformly distributed. In fact it is likely that the contrast
between dry and wet regions will increase as temperatures increase. This is shown
in Fig. 7.12, where one sees that the arid regions of the world, centered at latitudes
of 25° north and 25° south get 10–20 % drier. Alternatively, the polar regions and
a few spots in Africa receive significantly more precipitation.

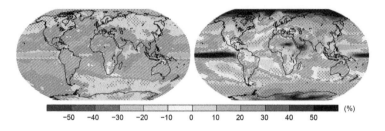

Figure 7.12: Change in average precipitation, 1986–2100. *Left*: RCP2.6. *Right*
RCP8.5.

Reproduced with permisson, IPCC AR5. See Fig. 7.10.

7.5.3   Changes in the Ocean

Water is much more capable of holding heat than air, so it may not surprise you
to learn that 90 % of the energy accumulated between 1971 and 2010 has been
stored in the ocean. The ocean will continue to warm during this century and heat
will penetrate from the surface to the deep ocean. The resulting thermal expansion
of the water and melting of the Arctic and Greenland ice sheets will contribute to
rising sea levels.

Figure 7.13 shows predicted sea level rises for the tree-hugger and business-
as-usual scenarios. Rises between 50 and 80 cm can be expected with 1 m not out
of the question. This will devastate low-lying areas around the world, the most
prominent of which is the Bangladeshi delta (the world's largest delta and one its
most fertile regions), that currently supports more than 100 million people.

Ex. Let's check this. The coefficient of thermal expansion for water
is around $2 \cdot 10^{-4}$ K^{-1}. The volume of the oceans is about

$$0.71 \cdot 4\pi R_{\oplus}^2 \cdot 4.2 \text{ km} = 1.5 \cdot 10^{18} \text{ m}^3. \tag{7.4}$$

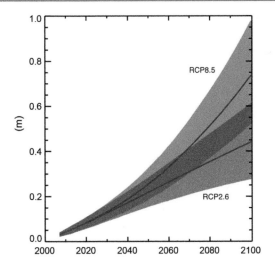

Figure 7.13: Predicted global mean sea level rise.

Reproduced with permission, IPCC AR5. See Fig. 7.10.

The numbers here are the fraction of the Earth covered by seas, the surface area of the Earth ($R_\oplus = 6.4 \cdot 10^6$ m), and the average depth of the oceans. A $4°$ increase in temperature increases this volume by 10^{15} m^3, which corresponds to a sea level rise of 3 m. This is in the right ball park. See Ex. 12 for why our result is too large.

Because carbon in water makes carbonic acid, the other great change to the ocean will be in its net acidity. Between 1995 and 2010 the acidity of the Pacific ocean has increased an alarming 6 %. IPCC predictions for the change in ocean pH are shown in Fig. 7.14. A historical drop from 8.15 to 8.10 is seen between 1950 and 2005. Since the pH is logarithmic, this corresponds to a 12 % increase in acidity. The pH is expected to reach 7.75 by 2100 in the RCP8.5 scenario, which is an increase in acidity by a factor of 2.5.

The impact of this increased acidity on marine organisms is not fully understood. The main concern seems to be for organisms that accumulate calcium in their tissues such as corals, molluscs, and crustaceans. There is already widespread concern over coral die-offs, although this is likely due to overfishing and mismanagement, rather than ocean acidification.

7.5.4 Other Changes

The implications of climate change that I have presented so far rest on well understood physical principles. Thus we are certain that the surface and ocean temper-

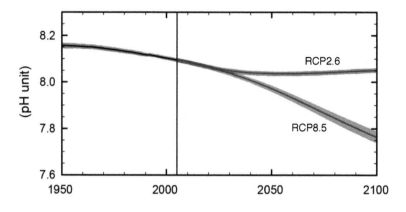

Figure 7.14: Past and predicted ocean surface pH.

Reproduced with permission, IPCC AR5. See Fig. 7.10.

atures will increase, that the sea level will rise, and that the acidity of the oceans will increase.[11] The only unknown is how much carbon we choose to burn.

Everything on the planet will be affected to some degree by the changes to come. Many of these changes are hard to quantify because they are incompletely understood. It is likely, for example, that glaciers and Arctic sea ice will continue to shrink. Similarly, permafrost will probably melt to some degree. This will release large quantities of methane, which is an effective greenhouse gas (see Table 7.2).

It is often stated that increased atmospheric carbon dioxide will help plant growth. But this is far from clear: stress due to increased temperature and shifting rain patterns can hurt. The IPCC has estimated the net result on some important crops, with results shown in Table 7.3. Most effects are small and quite uncertain. However, if wheat production really does fall at a rate of 2 % per decade, then

Table 7.3: Predicted crop yield change, by decade.

Crop	Median change	50 % confidence interval
Wheat	-2%	$-0.8 \rightarrow -3.5\%$
Soy	0	$-0.3 \rightarrow +1.1\%$
Rice	0	$0.0 \rightarrow -2.2\%$
Maize	-1%	$-0.6 \rightarrow -2.5\%$

Source: IPCC AR5.

[11] Virtually certain. It is possible, for example, that increased snow fall in the Antarctic will remove net water from the oceans.

the world will be producing 18 % less wheat by 2100. This must be made up by bringing more land into use, switching to other crops, or cultivating heat-resistant strains.

There are several climate system processes that could exhibit sudden changes due to increasing temperature. Examples are the Atlantic meridonal overturning circulation, Arctic sea ice, the Greenland ice sheet, and the Amazon forest. In principle, positive feedback in the climate could become so large that the temperature increase drives evaporation of the oceans. This situation is called a *runaway greenhouse effect* and is thought to have occurred on Venus, leaving that planet with a CO_2 atmosphere, sulphuric acid clouds, and a temperature of 460 C. Fortunately, the IPCC scientists regard all of these situations as unlikely to occur.

The IPCC has estimated that it will take a few hundred thousand years for the CO_2 we have emitted to leach out of the atmosphere. Thus our current actions constitute the greatest experiment with planet Earth ever undertaken by mankind. How do we respond? How do *you* respond? Maybe you think a few degrees of warming is nothing, especially compared to the prosperity that burning fossil fuels brings. Maybe you think we are driving the planet to ruin. We shall address these questions in Chap. 9.

REVIEW

Important terminology:

aerosol   [pg. 154]

albedo   [pg. 147]

anthropogenic carbon   [pg. 145]

carbon cycle   [pg. 145]

climate model qquad [pg. 153]

electronic, vibrational, rotational excitations   [pg. 149]

GtC and $GtCO_2$ [pg. 143]

greenhouse gas   [pg. 149]

insolation   [pg. 149]

Milanković cycle   [pg. 154]

proxy   [pg. 152]

radiative forcing   [pg. 150]

well-mixed gas   [pg. 152]

weather, climate   [pg. 153]

Important concepts:

Global energy budget.

Material and energy cycles.

Global carbon emissions amount to 9 pounds per person per day.

Positive and negative feedback.

FURTHER READING

L.R. Kump, J.F. Kasting, and R.G. Crane, *The Earth System*, Prentice Hall, 2010.

R. Pierrehumbert, *Principles of Planetary Climate*, Cambridge University Press, 2011.

G. Schmidt, and J. Wolfe, *Climate Change: Picturing the Science*, W.W. Norton & Company, 2009.

EXERCISES

1. US Carbon Footprint.

   Compute the carbon footprint of the average American.

2. Carbon Footprint of Gasoline.

   (a) If gasoline is primarily octane (C_8H_{18}), estimate the fraction of carbon in gasoline by weight.

   (b) Estimate the amount of carbon and CO_2 produced by burning a gallon of gasoline if the density of gasoline is 0.75 kg/L.

3. Gasoline Energy Content.

   If the primary process in burning gasoline is

   $$2\,C_8H_{18} + 25\,O_2 \rightarrow 16\,CO_2 + 18\,H_2O + \text{energy} \qquad (7.5)$$

   and burning a kilogram of gasoline releases $42\,\text{MJ}$ of energy, estimate the energy released in electron volts per reaction in the above equation.

4. Skin Temperature.

   Use the Stefan-Boltzmann law and a human power of $97\,\text{W}$ to estimate skin temperature at an ambient temperature of 23 degrees C.

5. Skin Temperature II.

   The previous question asks you to assume that human body heat is lost purely through thermal radiation. List ways in which this assumption is incorrect.

6. Skin Temperature III.

   A typical child has ½ the surface area of a typical adult. Assuming that a child's skin temperature is the same as an adult's, what is the average caloric intake for a child?

7. Thermal Radiation.

   We said that typical thermal energies are $0.01\,\text{eV}$. Verify this using the numerical value of Boltzmann's constant.

8. Clouds in the North.

   People who live in the north claim that if it is cloudy it will not be an exceptionally cold day. Is this a myth or is there a scientific reason for it?

9. Sailor Sayings.

   A saying amongst sailors is, "Red sky at night, sailor's delight. Red sky in morning, sailors take warning". Can you find a plausible reason for this?

10. Climate Model Grids.

    If the surface of the earth is divided into 120×120 squares,

    (a) How big are these squares in degrees?

    (b) How big are they in miles?

11. IPCC Scenarios.

    Which of the IPCC RCPs do you think is the most likely to occur? Why?

12. Sea Level Rise I.

See if you can find a reason that the predicted sea level rise in Sect. 7.5.3 was too large.

13. Sea Level Rise II.

Estimate sea level rise if 10 % of the Greenland ice sheet melts.

14. Denialism.

When the McCain-Lieberman bill proposing restrictions on greenhouse gases was being debated in the Senate on 28 July 2003, Senator James M. Inhofe made a two-hour speech in opposition. He cited a study by the Center for Energy and Economic Development in supporting his conclusion: "Wake up, America. With all the hysteria, all the fear, all the phony science, could it be that man-made global warming is the greatest hoax ever perpetrated on the American people? I believe it is."

Analyze Senator Inhofe's statement in light of Sect. 3.4.2.

15. The Solar Constant

You might have read that the energy that the sun supplies at the distance of the Earth, called the *solar constant*, is $1360\,\text{W/m}^2$. This is four times what was claimed in Sect. 7.3. See if you can find an explanation.

Nuclear Energy and Radiation

"If, as I have reason to believe, I have disintegrated the nucleus of the atom, this is of greater significance than the war."

— Ernest Rutherford, *in an apology for missing a meeting on antisubmarine warfare in 1917.*

"We knew the world would not be the same. A few people laughed, a few people cried, most people were silent. I remembered the line from the Hindu scripture, the Bhagavad-Gita... 'Now, I am become Death, the destroyer of worlds.'"

— J. Robert Oppenheimer, *upon testing the atomic bomb.*

Rutherford's simple experiment with a chunk of radioactive material and a piece of gold foil initiated an explosion of inquiry into the structure of the atomic nucleus. An early discovery was that the quantum world was far more complex than had been imagined. In 1932 three elementary particles were known, the electron, proton, and neutron. By the end of World War II dozens of particle were known and more were on the way. And of course, the new nuclear knowledge led to the construction of the atomic bomb and nuclear power reactors, while nuclear radiation became a bête noire.

In this chapter we will explore the development of nuclear physics, its applications to the bomb, nuclear power, and the health effects of nuclear radiation. We will also discuss the Chernobyl and Fukushima disasters, applications to medicine and food safety, and give a brief explanation of how stars shine.

8.1 Into the Atom

The tool that permitted Rutherford's discoveries was found by accident in 1896 by Antoine Henri Becquerel (1852–1908). Becquerel was a member of a distinguished family of scientists with a long interest in the phenomenon of

© Springer International Publishing Switzerland 2016
E.S. Swanson, *Science and Society*,
DOI 10.1007/978-3-319-21987-5_8

phosphorescence.[1] The recent discovery of X-rays motivated him to see if phosphorescent uranium salts also emitted X-rays. He tested this by exposing the salts to sunlight, wrapping them in black paper (to screen out the usual phosphorescence light), and placing them on photographic plates. When the plates were developed he found darkening, as one might expect from X-rays. But a few weeks later he decided to develop some plates even though it had been cloudy, and to his surprise found that the plates were still darkened. The conclusion had to be that uranium salts emit radiation (of some sort) naturally. Uranium is thus termed *radioactive*. Subsequent experiments revealed that the radiation bent in the presence of a magnetic field and hence was something new. Furthermore, when different substances were tested, some emitted radiation that did not bend (and so has no electric charge), while some had positive or negative charge. These classes of radiation were named ***alpha*** (positive), ***beta*** (negative), and ***gamma*** (neutral) by Rutherford.

Becquerel's discovery initiated a race to isolate and understand radioactive materials. Marie Curie (1867–1934) was the most successful to take up the challenge. Curie's life is a study in the vicissitudes of fortune. She was born as Maria Skłodowska in Russian-controlled Poland. Her mother died from tuberculosis when she was ten and her older sister died from typhus a few years later. Marie found it necessary to leave Poland at age 24 due to her involvement in a student revolutionary movement, and she ended up in Paris where she struggled to learn physics while trapped by poverty so severe that she often went without food. Life improved after she met a fellow physics student, Pierre Curie, and they married in 1895 (Fig. 8.1).

Within a year of Becquerel's discovery, Marie had discovered that thorium was radioactive, found the element polonium, and began a laborious effort to find the radioactive material in pitchblende. Eventually she isolated ½ gram of radium from one ton of pitchblende.

Pierre Curie died in 1906 after being run over by a carriage in the streets of Paris and Marie was offered his professorship at the University of Paris. In 1911 it was claimed that she had recently had an affair with another physicist, and the viscousness of the attack by the popular press forced her to hide out with her family.

During the First World War, Marie suspended her research and established a squadron of mobile X-ray facilities to help wounded soldiers. She eventually died of aplasmic anemia, likely brought on by a lifetime of exposure to nuclear radiation. She is the only person to win two Nobel Prizes in science and the first and only woman to be interred in the Pantheon in Paris. Upon her death Einstein said, "Of all celebrated beings, Marie Curie is the only one whom fame has not corrupted."

---

[1]A material that emits light after exposure to different frequency electromagnetic radiation is called phosphorescent.

Figure 8.1: Pierre and Marie Curie at their wedding.

8.2 Nuclear Decay

Rutherford's contributions to nuclear science overlapped those of Marie and Pierre Curie in their prime. Recall that Rutherford, Geiger, and Marsden discovered the nucleus and named the proton. Rutherford also suspected that another particle that was neutral and about the mass of the proton resided in the nucleus with protons. He studied Becquerel's radiation and soon noted that

(i) alpha rays are positive and can be stopped by a sheet of paper. They were subsequently identified as the nucleus of helium, namely two protons and two neutrons.

(ii) beta rays are negative and are more penetrating, requiring a thin sheet of aluminum to stop them. It was soon realized[2] that they are actually energetic electrons that are given off in a process that converts a neutron into a proton:

$$n \to p + e^- + \bar{\nu}_e. \tag{8.1}$$

---

[2]Becquerel measured the mass-to-charge ratio using Thomson's method and got the same value as the electron.

This process is called **beta decay**. The notation means that a neutron can decay into a proton, an electron, and a third particle, called an electron antineutrino.

(iii) gamma rays are neutral and very penetrating, requiring large of amounts of shielding. As we have seen, they are high energy photons.

The neutron was discovered by Rutherford's student, James Chadwick (1891–1974) in 1932. Chadwick had followed up earlier results from Germany that found an unusual radiation when polonium was used to irradiate beryllium. Chadwick suspected this was the neutron he and Rutherford had speculated about and performed a series of experiments to prove it.

We have reached the stage mentioned in the introduction. The microscopic world had a newly created and highly successful theory in quantum mechanics. The entities that played upon the quantum stage were electrons, neutrons, and protons. The atomic nucleus was made of positively charged protons and neutral neutrons. Electrons surrounded these at great distance in specific ways that explained the patterns seen in the periodic table. Within a few years the nature of the chemical bond had been understood by Linus Pauling and it looked as if a complete understanding of the microscopic – and by extension, the entire – universe was at hand, except for details.[3]

Recall that an atom can be in an excited state, where one or more electrons are promoted to higher orbitals. Because a nucleus is also subject to the rules of quantum mechanics, a nucleus can be in an excited state where a proton or neutron is promoted to a higher orbital. The lowest state – or **ground state** – can be specified by the number of protons and neutrons making up the nucleus. These are called Z and N respectively.[4] The total number of protons and neutrons is denoted A, which stands for the **mass number** familiar from chemistry. The more general term **nuclide** is used to refer to a specific combination of neutrons and protons with the notation

$$_Z^A\text{E},\tag{8.2}$$

where E is an element. Thus $_2^4\text{He}$ refers to a helium nucleus with two protons and two neutrons. By convention the element name is associated with the number of protons. An example of a nuclear beta decay shows this notation in use:

$$_6^{14}\text{C} \rightarrow {}_7^{14}\text{N} + e^- + \bar{\nu}_e.\tag{8.3}$$

Here carbon-14 decays to nitrogen-14.

---

[3] Very important details!

[4] Z stands for *Zahl*, which is German for number, and is traditionally used to denote the number of electrons (which equals the number of protons). Z is the **atomic number** of an element.

Recall that an alpha particle consists of two protons and two neutrons, thus an alpha decay is written (for example),

$$^{228}_{90}\text{Th} \rightarrow {}^{224}_{88}\text{Ra} + {}^4_2\text{He}. \tag{8.4}$$

Decays need not create nuclides in their ground state, for example molybdenum-99 can decay to technetium-99 in a nuclear excited state, which then emits a 140 keV photon and decays to its ground state.

$$^{99}_{42}\text{Mo} \rightarrow {}^{99}_{43}\text{Tc}' + e^- + \bar{\nu}_e \rightarrow {}^{99}_{43}\text{Tc} + \gamma(140\,\text{keV}) + e^- + \bar{\nu}_e. \tag{8.5}$$

It is possible that nuclides decay into nuclides that also decay. In fact, quite long decay chains are possible. An example that is important for nuclear power, called the thorium decay chain, is shown in Fig. 8.2. This chain starts with thorium-232 and proceeds via alpha and beta decay to lead-208, which terminates the chain because lead-208 is stable.

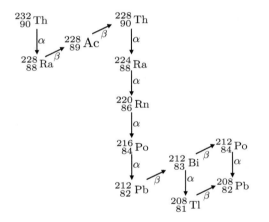

Figure 8.2: Thorium decay chain.

The quantum mechanical nature of the nucleus means that it is impossible to predict when a decay will happen. However, the probability of decay follows a definite curve, and for large times is a simple exponential. It is convenient to characterize the exponential with a *half life*, which is defined as the time it takes for one half of the available nuclides to decay on average.

Ex. Carbon-11 has a half life of 20 minutes. If you start with 1 kg of carbon-11 at noon, at 12:20 there would be ½ kg, and at 12:40 there would be ¼ kg remaining.

It is instructive to put all the nuclides into a chart as shown in Fig. 8.3. The x-axis is Z, the number of protons, and the y-axis is N, the number of neutrons. The black squares running down the middle of the figure are stable nuclides, starting with hydrogen-1 ($^1_1$H) at the lower left and ending with lead-208 ($^{208}_{82}$Pb) at the top right. The other squares represent unstable nuclides. Those near the stable region tend to have very long half lives (some of them are longer than the age of the universe), while those near the edge live for fractions of a second.

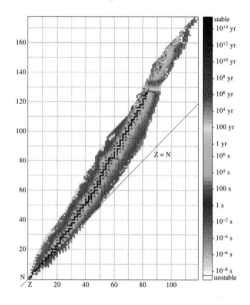

Figure 8.3: Chart of the Nuclides, showing lifetimes.

Q: Why does the stability region curve upwards?

A: The mutual electromagnetic repulsion between protons increases with Z so more neutrons are needed to make a stable nucleus.

The mode of decay shows some regularities in the plot. Almost all nuclides above the stability line decay by beta emission except for a few along the edge at the lower left, which decay by emitting a neutron (a fourth kind of nuclear decay). Those below the stability line decay by emitting a positron, which is the process

$$p \rightarrow n + e^+ + \nu_e. \tag{8.6}$$

A positron is identical to an electron but with the opposite charge.[5] The nuclides along the lower right edge can also decay via proton emission. Decay by alpha

---

[5]The positron is the **antiparticle** of the electron. It turns out all particles have paired antiparticles.

emission tends to occur at the upper end of the graph. Finally, there is small region centered at $Z = 100$ and $N = 150$ that decays via *fission*, which is the topic of the next section.

8.3 Nuclear Transmutation

Fission is a rare type of decay where a nucleus splits into two or more lighter nuclei. It was discovered just before World War II and, as you are probably aware, led very quickly to the creation of the atomic bomb and nuclear reactors.

The discovery of fission was made possible when scientists realized that the neutron would make a good tool for exploring the atomic nucleus. In fact, within a few years of the discovery of the neutron, Enrico Fermi (1901–1954) in Italy and Lise Meitner (1878–1968) and Otto Hahn (1879–1968) in Germany were using them to bombard uranium.

Q: Why is the neutron good for exploring the nucleus?

A: A neutral particle can approach many protons more easily than a positive particle.

Lise Meitner was an Austrian who had to overcome great difficulties to learn physics since women were not permitted into institutions of higher education. She eventually succeeded with the help of her parents, private tutoring, and her advisor Boltzmann. Upon arrival in Berlin, Max Planck permitted her to attend his lectures and hired her as a laboratory assistant. In 1912 she moved with Hahn to the Kaiser-Wilhelm Institute in Berlin and began her work in nuclear physics. This came to an abrupt end in 1938 when she had to flee Germany because of her Jewish heritage. She spent the rest of her life in Sweden and Britain, never returning to Germany.

Shortly after Meitner left Germany, Hahn and his assistant Fritz Strassmann (1902–1980) found barium (atomic number 58) after bombarding uranium (atomic number 92) with neutrons. This result baffled the scientists since it was expected that neutrons could only cause small changes in nuclei, not shatter them as their new result seemed to imply.

Hahn and Strassmann sent word of their discovery to Meitner, who pondered its significance. It was the Christmas holidays and Meitner discussed the findings with her nephew, Otto Frisch (1904–1979), who was visiting from Bohr's institute in Copenhagen. On a famous walk in the snow, the pair realized that the uranium nucleus must have been split by the bombarding neutron, creating nuclear fragments and some spare energy, which matched the 200 MeV of energy measured by Hahn and Strassmann. They called the new process *fission*, in analogy with the fissioning of cells. Fission reactions release a lot of energy that can be used for destructive or useful purposes. We will discuss these shortly, but first address the issue of the health effects of the new nuclear radiation.

8.4   Health Effects of Radioactivity

It is a radioactive world. Energetic particles bombard the Earth from outer space, radioactive rock is everywhere, even bananas are radioactive. The issue, of course, is *how* radioactive.

8.4.1   Measuring Radioactivity

There are several ways to quantify radioactivity:

becquerel (Bq) One decay per second. An older unit is the Curie, which is $3.7 \cdot 10^{10}$ Bq.

gray (Gy) One gray is equivalent to one joule of energy deposited in 1 kg of material.

sievert (Sv) The sievert is an attempt to quantify the effect of radiation on biological tissue. It is defined in terms of grays times a "quality factor". For example lungs have a quality factor of 0.12 and skin has a factor of 0.01. Older literature refers to "rem", which stands for "röntgen equivalent man". 100 rem = 1 Sv.

Obtaining a satisfactory estimate of risk is difficult. For example the gray only measures energy deposition in tissue. But we have learned in Chap. 5 that the energy of the radiation makes a difference – too low and its only effect is to heat tissue, not cause cancer or other health concerns. Chemical properties of the different elements can be important as well. For example, radium emits alpha particles that are very harmful, but only if the radium is inhaled. Similarly, iodine-131 concentrates in the thyroid gland where it emits beta and gamma radiation, while cesium-137 is water soluble and hence is easily excreted. The sievert, named after Swedish medical physicist Rolf Maximillian Sievert (1896–1966), is an attempt to average all of these effects into a useful measure of risk.

8.4.2   The Linear No-Threshold Model

It is generally agreed that a satisfactory model of the risk of developing cancer postulates a linear increase of cancer with dose at the rate of 5.5 % per sievert. The assumed linearity of this model has important consequences and is contentious if the linearity is assumed to continue down to very small doses. This assumption is called the *linear no-threshold model* (LNT).

The LNT model assumes further that the rate at which a dose is obtained is not important. Thus a 100 Sv dose obtained in 1 min is taken to be as dangerous as 0.5 Sv doses received daily for 200 days. In a similar fashion, if a dose of 1 Sv is assumed to give a 5.5 % chance of cancer, then a dose of 1/1000 Sv received by each of one thousand people will raise the average cancer rate in each person by 0.0055 %.

Ex. The average exposure to radiation in the United States is 6.24 mSv per year. The average life span is 77.4 years for males and 82.2 years for females. Thus the total exposure for men is 0.48 Sv and for women is 0.51 Sv. The lifetime risk of developing cancer in the USA is 43.9% for men and 38.0% for women. Since the LNT model equates 1 Sv with a 5.5% risk of developing cancer, we conclude that most cancer is not caused by nuclear radiation.

It is known that the LNT model is incorrect. Animals have evolved in a naturally radioactive environment and have developed cellular mechanisms to deal with damage to DNA. For example, some proteins crawl along DNA strands detecting damage, some proteins can repair damage, and others initiate cell death (*apoptosis*) if the DNA is beyond repair. Thus it is not sensible to assume that 200 1.0 mSv doses received over 1000 years will be as damaging as one 0.2 Sv dose received in a second. Indeed, there are scientists who believe that doses up to 100 mSv have no deleterious health effects at all.

This issue with the LNT model is contentious because it has important implications for public health policy. For example, if a 1000 Sv dose is accidentally released over a large area and affects 1,000,000 people, the LNT model predicts that each person will have a 0.0055 % additional chance of developing cancer, and that on average 55 people will develop cancer as a result of the accident. Alternatively, an epidemiologist who does not agree with the LNT model would argue that the 1.0 mSv dose received by each person is negligible and has no adverse public health outcome at all.

8.4.3 Background Radiation

The average radiation dose in the United States is 6.24 mSv per year. About one half of this is due to natural causes, chiefly radon, and about one half is due to medical procedures such as CT scans. Contributions from other sources such as atmospheric bomb tests and nuclear power are negligible. More details are given in Table 8.1

Table 8.1: Background radiation doses per person per year.

Source	USA dose (mSv)	World dose (mSv)
Radon	2.28	1.26
Food and water	0.28	0.29
Ground	0.21	0.48
Cosmic radiation	0.33	0.39
Medical	3.00	0.60
Cigarettes, air travel	0.13	–
Total	6.24	3.01

Sometimes natural sources of radiation can build up to more dangerous levels. For example radon is an odorless gas that occurs naturally as a decay product of uranium and thorium. The nuclide $^{222}_{86}$Rn has a half life of 3.8 days and decays via alpha radiation. Because of these properties radon can accumulate to unhealthy levels in basements. The highest radon concentrations in the United Sates are found in Iowa and Southeastern Pennsylvania.

Polonium is a dangerous radioactive element. In particular $^{210}_{84}$Po has a half life of 138 days and decays via alpha emission, which makes it exceptionally dangerous. The odds of dying from a 4.5 Sv dose is about 50 % – this dose can be delivered by only 10 ngm of inhaled polonium. Thus, in principle, one gram of polonium-210 could poison 20 million people and kill 10 million of them.[6]

Polonium achieved notoriety recently because it was used to murder Russian dissident spy Alexander Litvinenko in 2006. It is also likely that Irène Joliet-Curie, the daughter of Marie and Pierre Curie, died as a result from exposure to polonium. Ironically, polonium was discovered by Marie Curie and named in honor of her home country.

Polonium is also a public health concern because it is a decay product of radon, is produced by fertilizer, and tends to stick to tobacco leaves. Heavy smoking thus delivers a radiation dose of about 160 mSv/yr to the lungs which contributes substantially to lung cancer rates.

Background radiation dose is 3.1 mSv/yr.

8.4.4 Radiation Doses

Presumably one accepts the risk associated with the natural background radiation dose of 3.1 mSv/yr. In fact we are willing to double this risk in payment for the health benefits due to CT scans and other medical procedures. With this in mind, Table 8.2 shows a collection of doses that are associated with common things. Also shown are example doses for the nuclear accidents at Three Mile Island, Chernobyl, and Fukushima.

8.5 The Bomb

The fission process Hahn and Strassmann found corresponds to the scattering reaction

$$n + {}^{235}_{92}\text{U} \rightarrow {}^{236}_{92}\text{U}' \rightarrow {}^{141}_{56}\text{Ba} + {}^{92}_{36}\text{Kr} + n + n + n + \gamma \qquad (8.7)$$

Few who read and understood the reports of Hahn and Strassmann and Frisch and Meitner missed the significance of the result. Upon bombarding uranium with

---

[6]See Ex. 7 for more about the dosage.

Table 8.2: Typical radiation doses.

Source	Dose
One banana	100 nSv
Using a CRT monitor for one year	1 μSv
Dental X-ray	5–10 μSv
Flight from NY to LA	40 μSv
Living in a brick house for one year	70 μSv
Average dose for people within 10 miles of TMI	80 μSv
Dose at Fukushima town hall over 2 weeks	100 μSv
Mammogram	400–500 μSv
Average natural dose per year	3.1 mSv
Full body CT scan	10–30 mSv
Maximum annual dose for US radiation workers	50 mSv
Dose causing symptoms of radiation poisoning	400 mSv
50 % chance of dying	4.5 Sv
Fatal dose	8 Sv
Ten minutes next to the Chernobyl reactor after Accident	50 Sv

neutrons more neutrons are released along with substantial energy. This raised the possibility that a fission reaction could be *self-sustaining* under certain conditions, forming a *nuclear chain reaction*. And that meant a bomb of immense destructive power could be built.[7]

8.5.1 Little Boy

The prospect of the Germans building such a device jolted several people into action. In the US, Hungarian physicist Leó Szilárd (1898–1964) convinced his former teacher Albert Einstein to sign a letter warning President Roosevelt of the dangers of an atomic bomb. Meanwhile, Frisch and Rudolph Peierls (1907–1995)[8] wrote a famous memorandum that specified how a bomb could be built, estimated its destructiveness, and discussed the possibility that the Germans could build one.

These efforts led to the establishment of the Manhattan Project in 1942 with the goal of producing an atomic bomb. Frisch, Peierls, Fermi, and many British and refugee scientists eventually joined the project, helping to create the world's first atomic bombs.

---

[7]The last natural element in the chart of the nuclides is uranium, and only 1 % of that is uranium-235. Was nature trying to tease us?

[8]All four men were refugees from fascist Europe.

One of the main issues was the ***critical mass*** required to obtain a chain re-action. The problem is that a piece of uranium that is too small will permit the produced neutrons to fly out of the material before they can initiate other reactions. At some critical mass each reaction produces, on average, one additional reaction. Anything above this mass will explode with devastating effect. It is estimated that the critical mass of uranium-233 is 15 kg, uranium-235 is 52 kg, and californium-252 is 2.7 kg.

The second major issue concerned an idiosyncrasy with uranium. Natural uranium is more than 99 % $^{238}_{92}U$ and less than 1 % $^{235}_{92}U$. It might appear that there is not much difference between these, but uranium-238 simply absorbs extra neutrons and becomes uranium-239, while uranium-235 fissions. If there is too much uranium-238 present a bomb will fizzle out. The task, then, is to purify natural uranium until it is ***bomb grade*** – about 90 % uranium-235. This is a daunting undertaking because U-235 and U-238 are chemically identical. The current method is to put gaseous uranium in a ***centrifuge*** and spin it. This creates a strong artificial gravitational field in which the lighter U-235 floats to the center due to buoyancy. The inner lighter gas is then drained off into another centrifuge, and the process is repeated. Thousands of chained high precision, high speed centrifuges must work for years to produce enough U-235 to build a bomb.[9]

The "Little Boy" bomb was dropped on Hiroshima on August 6, 1945 (Fig. 8.4). It was made with 64 kg of uranium-235 that was brought to criticality by firing a doughnut shaped projectile at another piece in the shape of a spike. The subsequent explosion generated 67 terajoules of energy which destroyed much of the center of Hiroshima, and killed an estimated 66,000 people. This destruction was brought about by 3 years of industrial-scale effort, including building the towns of Los Alamos, NM; Hanford, WA; and Oak Ridge, TN.

The energy released by Little Boy is often given in terms of "tons of TNT" and comes to 16 kilotons. Within a few years the USA and the Soviet Union had developed "hydrogen bombs" with explosive power in the megaton range (the largest ever bomb exploded was around 50 megatons TNT equivalent). For comparison, 2.8 million tons of ordnance were dropped by the Allies in the European theater during WWII.

8.5.2   More Bombs

Since Little Boy was dropped over 2000 additional nuclear explosions have been set off for testing purposes. Five hundred of these were atmospheric tests, which released large amounts of radiation into the environment. The Centers for Disease Control and Prevention has estimated that this radiation will lead to 11,000 additional deaths from all forms of cancer. Should a government expose its own citizens to such risks?

---

[9]There is worry over Iran buying high speed centrifuges for enriching uranium. Are they attempting to build a nuclear bomb or merely enriching uranium for a domestic nuclear energy program?

Figure 8.4: Hiroshima, moments after Little Boy was dropped.

Public awareness of the issue was raised by the disastrous Castle Bravo hydrogen bomb test in 1954 on Bikini Atoll in the Marshall Islands. A wind shift and an underestimation of the power of the bomb led to the contamination of hundreds of people on neighboring islands. Atmospheric testing in Nevada lasted from 1951 to 1962 and has caused much concern for people living near the test site. As of 2014, 28,800 claims against the government have been approved with a total possible cost of $1.9 billion.

In 1962, Linus Pauling, who we have run into twice before, won the Nobel Peace Prize for his work to stop atmospheric testing of nuclear weapons. As a result, many countries ratified the Partial Test Ban Treaty prohibiting atmospheric nuclear testing in 1963. Subsequently, fear of the dangers posed by nuclear weapons has waned. The process was hastened by additional treaties reducing the number of warheads and the end of the Cold War. Consult Fig. 8.5 to judge for yourself whether the lack of concern over nuclear weapons is warranted.

The nuclear era had brought with it much cost in terms of human health, environmental impact, political destabilization, and a pervasive psychology of fear that gripped the world (see Fig. 8.6). The monetary cost was also not trivial; the Brookings Institute estimates that a staggering $5.8 trillion was spent on nuclear weapons between 1940 and 1996 ($7.6 trillion in 2014 dollars).

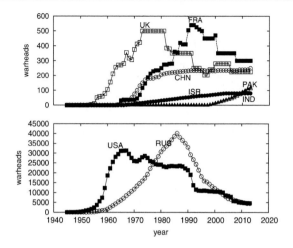

Figure 8.5: Global Nuclear Warheads. The figures do not include retired warheads.

Source: H.M. Kristensen and R.S. Norris, *Global nuclear weapons inventories, 1945–2013*, Bulletin of the Atomic Scientists, **69** 75 (2013).

Figure 8.6: Children were taught to "duck and cover" in case of nuclear attack.

8.5.3 Mass-Energy Conversion

Maybe you've heard that nuclear energy and the nuclear bomb are exemplars of Einstein's famous formula

$$E = mc^2. \tag{8.8}$$

Even famous physicists have made this connection in popular writing and TV programs. Sadly this is all nonsense. Einstein's formula simply says that mass is a form of energy. It is the zero velocity limit of the more general equation

$$E = \frac{mc^2}{\sqrt{1 - (v/c)^2}}, \tag{8.9}$$

which only becomes substantially different from the usual formula for kinetic energy ($+ mc^2$) at high speeds.

Recall that Frisch showed that fission products have a kinetic energy of about 200 MeV. The masses of these products is about $1/2 \cdot 235 \cdot 940$ MeV, which means that the speed of the fission products is about 6 % of the speed of light. This is a *low* speed, hence Einstein's formulas are irrelevant. What is in fact happening is that nuclear potential energy is converting into kinetic energy of the decay products in exactly the same way chemical reactions occur because chemical potential energy is converted into kinetic energy. The difference between nuclear and chemical reactions is only one of scale (Sect. 6.7).

8.6 Nuclear Power

Italian physicist Enrico Fermi (1901–1954) is the recognized father of nuclear power. Recall that we have seen him before as one of the people using neutron beams to explore nuclei. He thought that he had discovered **transuranics** (namely elements heavier than uranium) with his experiments, and duly won the 1938 Nobel Prize for it. This is likely the only prize awarded for incorrectly interpreted results since what Fermi actually found was fission. Fermi also made important contributions to quantum field theory, statistical mechanics, and developed the first theory that described beta decay. It is said, with good reason, that he was the last physicist who was adept at both theory and experiment.

Fermi's wife, Laura, was Jewish, and when they travelled to Sweden to receive his Nobel Prize they kept on going to New York, where they applied for permanent residency to escape Italian racial laws. This brought Fermi and his expertise into the American war effort where he worked on developing the "atomic pile".

8.6.1 CP-1

Fermi realized that controlling nuclear fission was the key to building a nuclear power reactor. One problem was that reactors tend to fizzle because neutrons from fission move too quickly to cause additional fission. Fermi dealt with this by inserting **moderators** into his reactor that slowed neutrons to more useful speeds. Water and carbon (in the form of graphite) make good and cheap moderators. In contrast to this problem, things could go catastrophically (i.e., explosively) wrong if too many slow neutrons were causing fission. The solution was to employ *control rods* that contain material that readily absorbs neutrons. Finally, heat generated by fission can then be used to boil water and run a steam turbine, just like in a regular power plant.

Fermi achieved all this by stacking graphite bricks and slugs of uranium in a wooden framework called "Chicago Pile 1" (CP-1). The control rods were made of cadmium and other metals that capture neutrons. There was no radiation shield and no cooling system. The resulting dramatic confirmation of Fermi's ideas was envisioned in a painting produced several years later (Fig. 8.7).

Figure 8.7: A depiction of CP-1 going critical under Stagg Field, University of Chicago, December 2, 1942. The volunteers in the back were meant to pour a neutron absorbing cadmium solution into the pile in case of emergency.

Reproduced with permission, Chicago History Museum, ICHi-33305, *Birth of the Atomic Age*, Gary Sheahan.

8.6.2   Modern Reactors

A wide variety of design strategies have been employed in modern nuclear reactors, and research continues to build safer, simpler, and cheaper models. There are six main reactor types in use around the world, differing in fuel enrichment level, moderator material, and coolant type. A breakdown of the world's reactors is given in Table 8.3. There are many more small research and military reactors as well.

Nuclear fuel is obtained from mined *yellowcake* (U_3O_8) which is converted to uranium hexaflouride gas (UF_6). The gas is then enriched to increase the uranium-235 content from 0.7 % to about 4 % and then converted into a hard ceramic oxide (UO_2) for use in the reactor. At this stage the uranium dioxide is only about 1000 times more radioactive than granite.

Water is an excellent moderator because hydrogen nuclei are nearly the same mass as neutrons and hence readily absorb neutron kinetic energy.

Ex. Imagine striking an object ball with a cue ball in snooker. If the object ball is hit squarely it will move off at the original speed of the cue ball, leaving the cue ball at rest.

Water also absorbs neutrons, which is overcome by enriching the fuel to 2–5 % uranium-235. Enrichment is expensive, but can be avoided by using **heavy water** instead of normal water as the moderator. In heavy water, the common hydrogen isotope is replaced by a heavier one called **deuterium**, which contains a neutron and a proton. Heavy water is much more efficient as a moderator because it absorbs very few neutrons, thus enabling natural uranium to be used as the reactor fuel. This strategy has been followed by Canada in the design of the CANDU reactor.

Table 8.3: Worldwide commercial nuclear reactors (2014)

Reactor type	Fuel	Moderator	Coolant	#
Pressurized water reactor (USA & France)	Enriched UO_2	Water	Water	250
Boiling water reactor (USA)	Enriched UO_2	Water	Water	58
Pressurized heavy water reactor (Canada)	Natural UO_2	Heavy water	Heavy water	48
Gas cooled reactor (UK)	Natural U, enriched UO_2	Graphite	Carbon dioxide	16
Light water graphite reactor (Russia)	Enriched UO_2	Graphite	Water	15
Fast breeder reactor (Russia)	PuO_2 & UO_2	None	Liquid sodium	2

Light water reactors typically extract 1 % of the energy available in their fuel because the fuel becomes **poisoned** by fission products that have built up to a level that renders the fuel inefficient. A way around this problem is to use **breeder reactors**, which are referred to in the last row of the table. As the name suggests, breeder reactors are capable of generating more fissile (fissionable) material than they consume. This material is typically plutonium-239. Because of this, breeder reactors can extract almost all of the energy available in nuclear fuel, which can decrease fuel requirement by a factor of 100. This may become a very important issue as resource availability dwindles in the future (see Chap. 9). Breeder reactors are not common today because they are more expensive to build, they generate plutonium (which can be used as nuclear bomb fuel), and because uranium is presently abundant.

Although uranium-235 is the only naturally occurring fissile material, it is possible to make nuclear reactors with other fuel. For example, thorium-232 is an abundant naturally occurring nuclide that can be converted into fissile uranium-233 via the process

$$n + {}^{232}_{90}\text{Th} \rightarrow {}^{233}_{90}\text{Th} \xrightarrow{\beta} {}^{233}_{91}\text{Pa} \xrightarrow{\beta} {}^{233}_{92}\text{U}. \qquad (8.10)$$

The reactor then runs as a breeder reactor with plutonium-239 replaced by uranium-233. This process is termed the *"thorium fuel cycle"* and has several advantages over the uranium fuel cycle including the relative abundance of thorium and the fact that dangerous plutonium-239 is not a product. In spite of this, current technology is so successful that there is little incentive to pursue the thorium fuel cycle option.

8.6.3   Waste Disposal

As fission occurs in a nuclear reactor, fuel bundles are contaminated with fission products that readily absorb neutrons. This eventually reaches a point where fission stops and the fuel bundle must be removed. Some countries **reprocess** used fuel bundles by dissolving them in acid and recovering the usable uranium and plutonium. Other countries, such as the United States, simply treat the used fuel bundles as waste. Unfortunately this waste is highly radioactive and must be carefully dealt with.[10] Spent nuclear fuel contains many fission products that fall near the middle of the chart of the nuclides and tend to decay rapidly.[11] Almost all fission products have half lives of less than 100 years. Transuranics, on the other hand, have very long half lives and after 1000 to 100,000 years generate most of the radioactivity of the spent fuel.

According to the International Atomic Energy Agency, the global nuclear industry generates about $10,000\,\text{m}^3$ of high level waste every year. This comes from about 30 tonnes of waste produced per typical 1000 MW nuclear power plant. For comparison, a typical 1000 MW coal plant produces 300,000 tonnes of ash per year, containing heavy metals, arsenic, mercury, uranium, and other toxins, which is dispersed in the atmosphere or buried in landfills.

When a fuel bundle is removed from a reactor it emits radiation that is lethal within a few minutes of exposure. After 100 years this level drops by a factor of 200, and after 500 years it drops by a factor of 60,000 to a level that is about 2000 times higher than the natural background level. This is considered safe for short exposures.

Altogether there is about 270,000 tonnes of used fuel in storage, much of it in on-site storage pools that provide shielding and remove excess heat. Although final disposal of this waste is not crucial at the moment, there is much debate about how to deal with it in the long run and just how safe pool storage is.

---

[10]To be sure, the waste from burning coal is also highly dangerous. The method for dealing with this is to disperse it through the atmosphere!

[11]Used fuel from light water reactors contains approximately 95.6 % uranium (less than 1 % is uranium-235), 2.9 % stable fission products, 0.9 % plutonium, and 0.6 % other fission products.

A major concern is that spent fuel contains much uranium and plutonium and it would be foolish to permanently lose this valuable resource. The US Department of Energy runs a ***geological repository*** (a fancy phrase for a hole in the ground) in Nevada called the Waste Isolation Pilot Plant (WIPP) where low and high level waste from the nation's nuclear weapons programs is being buried in chambers ½ mile below the surface. These chambers are cut into a bed of rock salt that is regarded as ideal for containing radioactive waste because it is free of subterranean water, easily mined, geologically stable, and fractures are self-sealing.

In 1982 the United States Congress passed the Nuclear Waste Policy Act which required the Department of Energy to build and operate a geological repository for the nation's commercial radioactive waste. In 1987 Congress selected the Yucca Mountain range in the Nevada Test Site of Southwest Nevada as the repository location.[12] By 2008 Yucca Mountain had become the most studied piece of geology in the world, with some $9 billion spent on geological and materials science studies. These funds came from a special tax on electricity that was paid by the consumer. Political wrangling over the site was intense with opposition objections ranging over the possibility that the land was sacred to Native Americans, dangers of transporting fuel over the nation's roads and rail lines, fear of water seepage or earthquakes, and a sense that Nevada was the nation's dumping ground. The project sustained a major blow in 2004 when the Court of Appeals of the District of Columbia ruled that the Department of Energy had to guarantee maximum radiation levels for a period of one million years, rather than the generally accepted 10,000 years. In the end, funding was terminated by Energy Secretary Steven Chu in 2011. The nation currently has no plans for disposing of nuclear waste in a sensible fashion.

It is of course ludicrous to expect that anything can be projected one million years into the future. Furthermore, permanent storage should not be considered in the first place since spent fuel is a valuable resource and should be kept in recoverable storage for future generations. Meanwhile, billions of dollars of taxpayer money has been squandered and development of one of the few viable carbon-free sources of energy we have has been hindered.

8.7 Nuclear Disasters

Nuclear radiation is dangerous in proximity to people. As we have seen, the primary safety mechanisms in the nuclear industry rely on keeping radioactive material away from people. This methodology is essentially foolproof until a greater fool comes along. We review three major disasters in the commercial nuclear industry in this section. This is partly a study in failure modes of complex systems, but here we are mainly interested in the health risks posed to the public when things go very wrong. As usual, it is important to set scales for the purposes of making comparisons. We will therefore examine health risks due to the

---

[12]This means that the site selection was political rather than scientific.

coal industry in Chap. 9 (in this case risks are associated with normal operating procedures, not disasters).

All three disasters followed a too-typical pattern, beginning with minor technical issues that were compounded by poor information and poor decision making leading to situations that rapidly spiralled out of control. This is representative of a system with positive feedback (see Sect. 7.1.2), where a small perturbation results in an effect that also perturbs the system in the same direction. Inherently safe systems feature negative feedback, where small perturbations led to effects that tend to damp out the initial perturbation. Unfortunately, large systems like nuclear reactors are sufficiently complex that it is very difficult to design them to feature negative feedback under all conditions – especially when ill-advised human intervention is present.

8.7.1 Three Mile Island

On March 28, 1979 reactor 2 of the Three Mile Island nuclear power plant near Harrisburg, Pennsylvania experienced a catastrophic partial meltdown and release of radioactive material. The accident started when technicians followed an unusual procedure in attempting to clean a water filter. This forced a small amount of water into an air line, which led to a shutdown of the system feeding water to the steam generators. The sudden lack of a heat sink caused the reactor core to overheat and initiated an automatic emergency shutdown where control rods are inserted into the reactor.

The story should have stopped here with an expensive, unnecessary, but safe reactor shutdown. Unfortunately another human error led to disaster. Although the reactor core was shut down it was still generating heat from residual radioactive decays. Since this heat was not being carried away by the steam turbine system, a backup water coolant system was supposed to take over. But this had been shut down for maintenance and the reactor overheated,[13] causing a steam pressure relief valve to trip. This valve worked as required, but then stuck in the open position allowing coolant water to escape the reactor and creating a positive feedback loop.

At this stage yet more human error entered the picture. Poor interface design and training led the operators to assume that *too much* coolant was in the system and they turned off the emergency core cooling pumps. The resulting loss of coolant and higher temperatures formed steam throughout the cooling system, effectively shutting it down. About two hours after the start of the incident the reactor fuel elements were melting and a potentially explosive hydrogen bubble had formed. A state of emergency was declared an hour later.

Seemingly minor circumstances had led to disaster. Nevertheless, the containment building housing the reactor did not suffer damage and the released radioactivity caused minimal damage. It was estimated that 93 PBq of radioactive noble

---

[13] Running the reactor with the backup system down was against regulations.

gases were released. This stupendous radioactivity was dispersed over a wide area, with an average dose of $14\,\mu Sv$ to 2 million people living near the plant (see Ex. 3 for an estimate of the resultant cancer incidence). Direct measurements of radioactivity found no increased levels and subsequent epidemiological studies found no increased cancer rates.

8.7.2 Chernobyl

Unfortunately, the relatively benign outcome of the TMI accident was not repeated 7 years later when a Russian graphite water reactor experienced a catastrophic meltdown. This occurred at reactor 4 of the Chernobyl Nuclear Power Plant in the Ukraine in an event universally known as "Chernobyl".

The accident had its roots in an ill-advised experiment to see if a steam turbine could generate power for a short period after a shutdown. The experiment protocol called for reducing reactor 4's operating power from 3200 to 700 MW. However, a technician pushed control rods too far into the reactor, lowering the power output to 30 MW, which was too low for a safe test. Technicians then overrode the automatic control rod system and manually withdrew several rods, which led to rapid poisoning and unstable conditions in the reactor. The reactor eventually settled down at 200 MW power and it was decided to proceed to the next step in the experiment, which was increasing the coolant flow into the reactor. This reduced reactor power, which led operators to decrease the coolant flow and withdraw more control rods.

In the final phase of the experiment, the feed to the steam turbines was shut off. The resultant lack of a heat sink raised the reactor core temperature, creating steam bubbles, which reduced the ability of water to absorb neutrons, which further raised temperatures. The automatic control rod system was able to counter most of this positive feedback loop, but most of the rods had been placed under manual control. Presumably the feedback overwhelmed the control system, because at this point the emergency reactor shutdown button was pressed, sending all control rods into the reactor. In a design flaw, this displaced moderating water for a few seconds, which caused another hike in the reactor temperature, jamming the control rods. The reactor soon ran out of control. The resulting explosion destroyed the reactor housing and the surrounding graphite moderator caught fire. Vast amounts of radiation were released to the immediate vicinity because it was not thought necessary to put the reactor in a containment building.

Thirty-one people died as direct result of the catastrophe, including three volunteers who went on a suicide mission to open vital sluice gates to permit a pool of water below the burning reactor to be drained.

The Chernobyl explosion put 12 EBq of radioactive material (about 400 Little Boys) into the Earth's atmosphere. As a result, 350,000 people were evacuated from the most severely contaminated areas of Belarus, Russia, and Ukraine between 1986 and 2000. The radioactive plume from the fire drifted over large parts western Europe causing great concern.

A UN agency has estimated a global collective dose of radiation exposure from the accident "equivalent on average to 21 additional days of world exposure to natural background radiation" (see Ex. 4). The ½ million recovery workers received far higher doses, amounting to an extra 50 years of background radiation each.

Estimates of the number of deaths due to the disaster vary widely. In 2005 the World Health Organization estimated that 4000 additional deaths due to cancer amongst the emergency workers and evacuees will result. Alternatively, the Union of Concerned Scientists used the LNT model to obtain a figure of 25,000 excess cancer deaths.

8.7.3 Fukushima

On March 11, 2011 the fifth most powerful earthquake in recorded history occurred 70 km off the coast of Japan. The resulting tsunami surmounted a seawall protecting the six BWR reactors of the Fukushima Daiichi nuclear power plant. All functioning reactors initiated emergency shutdowns, but as we have seen, residual heat is still produced and needs to be removed. This, however, did not happen because the tsunami flooded the diesel generators powering the emergency pumps. A back-up battery system gave out the next day. The subsequent increase in temperature caused the release of hydrogen gas, which led to several explosions over the next 3 days as the cores of three reactors suffered partial meltdown.

It is estimated that 27 PBq of cesium-137 (with a 30 year half life) was released into the ocean between March and July of 2011. The overall release was about 1/10 that of Chernobyl.

No immediate deaths resulted from the disaster, although two workers were treated for radiation poisoning. Approximately 300,000 people were evacuated from the neighboring area. Radiation exposure was measurable but not dangerous. In August 2012, researchers found that 10,000 nearby residents had been exposed to less than 1 millisievert of radiation, significantly less than Chernobyl residents. In the two most affected locations of Fukushima prefecture the preliminary estimated radiation effective doses for the first year ranged from 12 to 25 mSv. And the World Health Organization indicated that evacuees were exposed to so little radiation that radiation-induced health impacts are likely to be below detectable levels.

Of course reassurances such as these do little to assuage fear over nuclear catastrophe in the public mind. There is currently much worry over the effect of the radiation on the west coast of America. Indeed, 27 PBq is a lot of radioactivity, but the Pacific Ocean is big. Simulations of ocean currents indicate that the concentrations of cesium-137 on the west coast will not surpass $20\,\text{Bq/m}^3$. This is high compared to natural levels of cesium in the Pacific (see Table 8.4), but poses no health risk.

Table 8.4: Concentrations of cesium-137 in the world's oceans.

Body of water	Activity of cesium-137 (Bq/m^3)
Pacific Ocean	1.8
Atlantic Ocean	1.0
Indian Ocean	1.2
Black Sea	16
Baltic Sea	40
Irish Sea	61

Source: Woods Hole Oceanographic Institute.

8.8 Other Uses of Radioactivity

The headline applications of nuclear radioactivity have been nuclear power and bombs, but more mundane uses exist. In fact it was not long after the Curie's discovery of radium in 1898 that it was being sold by medical quacks as a cure-all. This is despite of an experiment which Marie Curie performed on herself where she carried a small sample of radium next to her skin for ten hours. The result was an ulcer that appeared within a few days. Radium is luminescent and a more legitimate use was to paint glow-at-night watch dials. The subsequent scandal over induced cancer in the "Radium Girls" who painted the dials helped establish worker rights legislation in the United States.

8.8.1 Smoke Detectors

Smoke detectors are probably the most common point of contact with nuclear technology. Smoke detectors contain tiny ($0.25\,\mu\text{g}$) amounts of the radioactive isotope americium-241 in an ionizing chamber that is capable of detecting smoke. Americium-241, $^{241}_{95}\text{Am}$, has a half life of 432 years and is an alpha emitter. It is produced in nuclear reactors in the uranium-238 decay chain. Emitted alpha particles ionize air molecules that carry a small current across the ionizing chamber in the smoke detector. Any smoke particles that are present in the chamber bind to the ions and cause the current to drop, which sounds an alarm.

8.8.2 Medical Isotopes

Various exotic isotopes have found use in the medical industry. Amongst these are several isotopes that are used to kill cancer cells with nuclear radiation in a rather brute force manner. The chemical properties of the isotopes are matched to the desired cancer. For example, iodine is taken up by the thyroid gland while strontium accumulates in bone.

iodine-131 treating thyroid cancer (half life: 8 days)

strontium-89 treating bone cancer (half life: 52 days)

yttrium-90 treating lymphoma (half life: 2.7 days)

cobalt-60 treating skin, prostate, cervical, breast cancer, and tumors (half life: 5.3 years)

cesium-137 similar to cobalt-60 (half life: 30 years)

Radioactive isotopes also find use in sophisticated diagnostic techniques. A leading example is ***positron emission tomography***, commonly called a PET scan. This diagnostic relies on the emission of positrons from the decay of flourine-18 ($^{18}_{9}Fl$) which has a half life of 110 min (see Eq. 8.6 to remind yourself of inverse beta decay). Positrons can ***annihilate*** with electrons to produce a pair of high energy photons that can be detected outside the body. The trick is to bind flourine-18 with a biomolecule to carry it to useful places. The most commonly used compound is fluorodeoxyglucose, which is metabolized by the body because it is a sugar. This is useful in cancer diagnosis because cancer cells reproduce rapidly and therefore make high demands on the metabolic system, leading to accumulation of the biomarker in tumors.

A similar technology is based on the nuclide technetium-99m, which is a long-lived excited state of technetium-99. The "m" denotes metastable and indicates that this substance is an excited state of technetium-99. We have used the notation $^{99}_{43}Tc'$ before, but stick to common usage here. Technetium-99m has a half life of 6 h and decays via the emission of a 141 keV gamma ray that can be imaged by a SPECT (single photon emission computed tomography) camera. Typical doses are 1 GBq which results in an exposure of about 10 mSv to the patient – about three times the average annual dose. It is estimated that this gives rise to one case of cancer per 1000 patients. Since 20 million diagnostic tests are administered every year, about 20,000 additional cases of cancer occur due to this relatively high dosage test. Of course, this number needs to be balanced against the lives saved due to the test's diagnostic capabilities. Furthermore, we have used the problematic LNT model to make this prediction.

You might wonder how a substance with a half life of six hours can possibly be produced, shipped, and dispensed before it disappears. It turns out that technetium-99m is a decay product of another nuclide called molybdenum-99, which has a longer half life of 66 h. This material is constantly produced in (only!) five reactors around the world and is shipped in ***generators*** to hospitals where the technetium-99m can be extracted for use.

8.8.3 Thermoelectric Generators

Alpha decays of radioisotopes are effective at heating surrounding material. This heat can be converted into electricity with thermocouples and thus devices, called *radioisotope thermoelectric generators* (RTGs) can be used as batteries for applications that require low power (typically 100 W) for long periods of time. An RTG using plutonium-238 is currently powering the Cassini mission to Saturn and powered the Galileo mission to Jupiter before it was terminated by plunging into Jupiter's atmosphere in 2003.

8.8.4 Food Irradiation

Exposing foodstuffs to ionizing radiation kills bacteria and therefore can be an important tool in the era of industrialized food production. A source of such radiation is gamma rays produced from the decay of a radioisotope like cobalt-60 (half life: 5.3 years). Unfortunately, the public perception of radioactivity is so negative that irradiated food is rarely seen, despite its obvious public health benefits.

One issue is the idea that irradiating food makes it radioactive. This is implausible because biologically damaging ionizing radiation has energies of 10s of eV, while the energy required for nuclear transmutation is in the 10s of MeV (see Sect. 6.7). However, this simple argument fails in the case of cobalt-60 because this isotope emits very energetic 1 MeV gamma rays. If this energy were in neutrons instead of photons, it would be sufficient to initiate fission in uranium-238.[14] In practice, these gamma rays do not induce any radioactivity in irradiated material.

The other cause for public concern is that irradiation "does something" to food. Indeed, along with disrupting molecules in bacteria, energetic photons also disrupt molecules in the surrounding food. However, careful study shows that this disruption is substantially milder than that induced by standard food processing such as cooking, salting, or curing.

When public fear is combined with craven politicians, toxic and foolish policy can arise with surprising ease, as the following example shows.

> Ex. Australian Ban of Irradiated Cat Food
>
> In 2008, a series of cat deaths in Australia was blamed on irradiated cat food sold by Champion Petfoods. Despite the fact that all cat deaths were associated with a single batch of cat food, and in spite of the lack of evidence concerning irradiation, irradiating cat food was banned in Australia in 2009.
>
> This action was clearly taken to appease a worried public. Surely if irradiated food is unhealthy for one large mammal, it is unhealthy for

---

[14]Which, of course, should not be present in your food!

all large mammals. Apparently this has occurred to no one as the irradiation of dog food remains legal – with the exception that it must be labelled "Must not be fed to cats."

8.9   Why Stars Shine

As we have seen, unlocking the atom's secrets led directly to new and powerful knowledge about the workings of nature. Once the right ideas were in place, progress was rapid on many fronts, including the resolution of a longstanding controversy about the source of the sun's energy.

8.9.1   A Fire in the Sky

The ancient Greeks imagined the sun to be a fire in the sky, and this view held for thousands of years until science finally caught up with the issue. The most powerful voice in the debate was that of William Thomson (1824–1907), 1st Baron Kelvin, commonly known as Lord Kelvin. Kelvin was an Irish prodigy who became a professor at age 22, grew wealthy laying the first transatlantic telegraph cable, and made fundamental contributions to thermodynamics. Kelvin (along with Helmholtz) was convinced that the sun's luminosity was produced by the conversion of gravitational energy into heat, likely due to infalling meteors.

In 1862 he declared in a public lecture:

> "That some form of the meteoric theory is certainly the true and complete explanation of solar heat can scarcely be doubted, when the following reasons are considered:
>
> (1) No other natural explanation, except by chemical action, can be conceived.
>
> (2) The chemical theory is quite insufficient, because the most energetic chemical action we know, taking place between substances amounting to the whole sun's mass, would only generate about 3,000 years' heat.
>
> (3) There is no difficulty in accounting for 20,000,000 years' heat by the meteoric theory."

Source: "Popular Lectures and Addresses", Baron William Thomson Kelvin.

Notice that Kelvin has dismissed the old Greek idea: a chemical fire would simply burn out too fast. Seeing no other source of energy, he assumed it comes from gravitation via infalling meteors, and computed an age of 20 million years for the sun (which sets an upper limit on the age of the Earth). This estimate was much shorter than that made by Darwin in support of his theory of evolution. In fact, Kelvin's reputation was so fearsome that Darwin stopped making references to the age of the Earth.

Kelvin's argument held sway until Becquerel made his discovery. It was soon realized that radioactive materials release heat and not long after that Rutherford and others suggested that radioactivity might be the source of the sun's energy. But observations of the sun's spectral lines revealed no radioactive elements – in fact the sun appeared to be made up almost entirely of gaseous hydrogen. Something else was needed.

A step closer to the truth was taken in 1920 by the brilliant English astrophysicist, Sir Arthur Eddington (1882–1944), who noted that four hydrogen atoms could convert to a helium atom with an excess of energy that could power the sun for 100 billion years. Eddington gave a public address about his ideas during which he made a remarkably insightful observation

> "If, indeed, the sub-atomic energy in the stars is being freely used to maintain their great furnaces, it seems to bring a little nearer to fulfillment our dream of controlling this latent power for the well-being of the human race – or for its suicide."

Eddington had no detailed idea about how his proposed mechanism worked. But he was on the right track, and from here on things progressed quickly.

8.9.2 Alpha Decay

The next breakthrough came as a result of solving another longstanding problem. Alpha decay had been known and used as a tool for 30 years, but it was not understood until George Gamow (1904–1968), Ronald Gurney (1898–1953), and Edward Condon (1902–1974)[15] realized that it could be explained with the new quantum mechanics.

Alpha decay was an utter mystery before the advent of quantum mechanics. How could something be stable and then spontaneously turn into two other things? Worse, it seemed to happen randomly! Our trio realized that some of the famous properties of quantum mechanics provided a likely explanation. In Chap. 6 we saw that the quantum nature of the atom provides a rigidity that prevents the atoms from collapsing. The wave nature of particles and the mathematics of quantum physics also implies that deterministic statements about the microscopic world cannot be made – only probabilities of events can be computed. This fit perfectly with the random nature of alpha decays. But why did the decays happen?

---

[15] All of these men were distinguished physicists. It would take us too far afield to describe their careers, but I cannot resist noting that Condon was denounced by Hoover as a communist spy, and viciously attacked during the McCarthy era. One assailant said, "You have been at the forefront of a revolutionary movement in physics called . . . quantum mechanics. It strikes this hearing that if you could be at the forefront of one revolutionary movement . . . you could be at the forefront of another."

Imagine that you want to roll a ball from a valley in which you stand over a hill and into the next valley. If the hill is not too high you can throw the ball with enough energy that it will reach the top of the hill and roll down into the next valley. If the hill is too high you will not be able to reach the top with the ball. For some reason private to you, you *really* want to get the ball over the hill. Your frustration grows because you know that if you could just reach the top of the hill the ball would pick up energy rolling down the other side. If only you could borrow energy from nature and pay it back later! Alternatively, if you could just run a tunnel through the hill, your ball dreams could be easily achieved.

The breakthrough for alpha decay came when Gamow and colleagues realized that the indeterminant nature of quantum mechanics allows exactly this kind of "borrowing". In analogy with the ball and hill scenario, the concept is called *quantum tunneling*. In the case of alpha decay, the alpha particle is held in the nucleus by a force, but the force gets progressively weaker as the distance between the alpha particle and the rest of the nucleus increases. Eventually it becomes repulsive. This is analogous to our hill: before the top of the hill gravity pulls the ball back; past the top of the hill gravity pushes the ball away (or rather, pulls it into the next valley). The wave nature of the alpha particle means that, although it spends its time on the attractive side of the force, it can spontaneously appear on the repulsive side. Once it is there, it is driven away from the rest of the nucleus and the decay occurs.

You might be curious about what "causes" the decay. After all, energy is conserved so there is no benefit to the alpha particle breaking away from its parent nucleus. The answer is that the decay process increases the entropy of the universe: the disorder in the system increases and potential energy is turned into kinetic energy.

8.9.3 Powering the Sun

Within a year a young Dutch-German experimental physicist named Fritz Houtermans (1903–1966) realized that the same effect could happen in reverse, a process called nuclear *fusion*. Houtermans was both one quarter Jewish and a communist, and found it prudent to leave Germany in the run up to World War II. He first went to England, but could not tolerate the conditions there.[16] Houtermans left England for the Soviet Union in 1934. His research and life went well at first, but by 1937 the worsening situation in Europe and Stalin's paranoia caught up with him, and he was arrested and tortured by the secret police. It is difficult to believe that things could get worse, but in 1940 he was extradited into the hands of the Gestapo. Somehow he survived both imprisonments and the war.

Let us go back to 1928 when Houtermans had realized that quantum tunneling meant that nuclear fusion could take place. It turns out that the process is too

---

[16]He said, "the limits of the Roman Empire could be inferred from the way in which potatoes are prepared: if they were just boiled with salt you were certainly beyond those limits." England was assigned to "the domain of salted potatoes".

rare to be detectible in laboratories, but it could take place in the infernos inside of stars. Houtermans and a young British physicist, Arthur Atkinson, therefore wrote a paper putting forward the radical idea that nuclear fusion is the source of the sun's energy. The feeling of discovery must have been remarkable:

> "That evening, after we had finished our essay, I went for a walk with a pretty girl. As soon as it grew dark the stars came out, one after another, in all their splendour. "Don't they shine beautifully?" cried my companion. But I simply stuck my chest out and said proudly: "I've known since yesterday why it is that they shine."

Houtermans' idea was picked up nine years later by yet another refugee scientist named Hans Bethe (1906–2005), who developed it into a detailed theory that earned him the Nobel Prize. Bethe's brilliance and initiative placed him at the leading edge of physics for a substantial portion of the twentieth century. As a result his name bedecks dozens of theorems, equations, and ideas.

Bethe used the idea of Houtermans and the equations of alpha decay to make a systemic examination of all possible nuclear fusion processes in the sun. The rate at which these reactions occur depends on the temperature. Based on this, he concluded that the *CNO cycle* powers the sun. This is a catalytic process in which carbon, nitrogen, and oxygen assist in converting four protons into an alpha particle, two positrons, and two electron antineutrinos, as shown in Fig. 8.8. The alpha particle is written as ^{4}He and β^+ denotes a positron and an electron antineutrino. Notice that two protons are converted into neutrons, which is why the positrons and antineutrinos must be present. Also 25 MeV of energy is liberated in photons. It is this energy that powers the sun and all life on Earth.

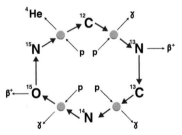

Figure 8.8: The CNO cycle. Four protons are converted to photons, two positrons, two neutrinos, an alpha particle, and 25 MeV of energy.

It turns out that Bethe was wrong about the importance of the CNO cycle in the sun. It is strongly temperature dependent and actually dominates in stars that are slightly heavier than the sun.

The dominant process in the sun is called the *proton-proton chain reaction* (it is not a chain reaction in the sense of nuclear energy), and was also studied by

Bethe. The initial part of the reaction (see Fig. 8.9) involves two protons fusing into a **deuteron**, which is a bound state of a proton and a neutron and is part of heavy water discussed previously. This requires converting a proton into a neutron via inverse beta decay, shown in Eq. 8.6. Although the neutron weighs more than the proton, the deuteron is lighter than a proton and a neutron; and is also lighter than two protons, thus the deuteron is stable. However, the process does not stop here because a deuteron can combine with a proton to make helium-3 (two protons and a neutron). Two helium-3 nuclei can combine to produce two free protons and a helium nucleus. At this stage the process stops, having converted four protons into a helium nucleus, two positrons, two antineutrinos, and 26 MeV of energy in the form of photons.

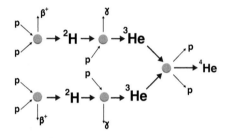

Figure 8.9: The pp Chain Reaction. Four protons are converted to photons, two positrons, two neutrinos, an alpha particle, and 26 MeV of energy.

Converting two protons into a deuteron is a rare process because inverse beta decay is itself very rare. For this reason the sun has illuminated the Earth for 4.5 billion years and will continue to do so for another 5.5 billion years.

Ex. Computing the life span of the Sun.

The life span of the sun can be computed by dividing the mass of hydrogen in the sun by the rate at which it is being burned. We assume the burn rate is equal to the rate the sun is emitting energy and that this is equal to its luminosity.

The sun's luminosity at Earth is 1357 W/m^2. Thus the total luminosity is $1357 \cdot 4\pi R^2$, where $R = 1.5 \cdot 10^{11}$ m, which is $3.8 \cdot 10^{26}$ W. Each proton-proton chain reaction liberates 26 MeV = $4.2 \cdot 10^{-12}$ joules of energy. We neglect the contribution of neutrinos, so this energy goes into photons. A watt is a joule per second so there are $9.0 \cdot 10^{37}$ pp reactions per second burning four protons of mass $6.6 \cdot 10^{-27}$ kg. The pp process happens in the core of the sun, which comprises about 10% of the total mass of $2 \cdot 10^{30}$ kg. We obtain a burn rate of $6 \cdot 10^{11}$ kg/s. Finally, dividing this into the hydrogen mass gives a total lifetime of $\frac{1}{3} \cdot 10^{18}$ s = 10 billion years.

8.9.4   Nucleosynthesis

Bethe's solar model could only explain the production of helium; the explanation of the creation of all the heavier elements had to await the generation of physicists who worked after World War Two. Leading the charge was Sir Fred Hoyle (1915–2001), a British astronomer who liked to write science fiction and stir controversy with unorthodox ideas about evolution and interstellar viruses.

In 1946, when Hoyle still practiced conventional physics, he showed that stars can become very hot as they age, and that these nuclear furnaces are capable of synthesizing heavy elements. Five years later he showed that the elements between carbon and iron arise from supernovas.

The theory was developed in detail and could soon explain the abundances of the various elements in the universe and explain how more complex elements appear as the universe aged. Thus almost all the elements on earth were formed in the distant past in stellar interiors. As Carl Sagan was fond of saying,

> We are made of star stuff.

The exceptions to this neat picture were hydrogen and helium, whose abundances (and existence) remained unexplained. It is clear that one cannot explain the existence of everything – something must have been there to define the beginning. And in fact, it is to the beginning of the universe that we must look for answers.

Scientists have known since the 1920s that the universe is expanding. Of course this means that the universe must get smaller as one goes backwards in time. It is tempting to extrapolate back to when its size was zero, some 14 billion years ago, and assume that the universe started then in some sort of hot quantum explosion. This concept was referred to as the "Big Bang" by Fred Hoyle in 1949 and has been known by that name ever since.

George Gamow and his PhD student Ralph Alpher (1921–2007) were the first to examine the implications of element creation in an expanding and hot universe.[17] They assumed that the universe began as a hot and dense soup of neutrons and that nuclei are created as they successively capture neutrons and protons (protons arise as neutrons beta decay). This process continues until the universe expands and cools enough to shut the nuclear reactions down. The concept came to be known as "Big Bang nucleosynthesis" and has been corrected and modified since its introduction in 1948. In its present form, the model explains the abundances of deuterium, helium-3, and helium-4 due to nucleosynthesis in the first three minutes of the universe's existence.

---

[17]Curiously, the paper reporting the ideas of Alpher and Gamow includes Hans Bethe in the author list. Gamow was famed for his quirky sense of humor, and noting that the first three letters of the Greek alphabet are alpha, beta, and gamma, he decided to simply add Bethe's name.

REVIEW

Important terminology:

A: number of protons and neutrons

alpha, beta, gamma radiation   [pg. 166]

becquerel, gray, sievert   [pg. 172]

breeder reactor   [pg. 181]

chart of the nuclides   [pg. 170]

CNO cycle   [pg. 193]

critical mass   [pg. 176]

decay chain   [pg. 169]

deuterium   [pg. 181]

fission   [pg. 171]

fusion   [pg. 192]

half life   [pg. 169]

moderator   [pg. 180]

nuclear waste repository   [pg. 183]

nuclide, isotope   [pg. 168]

positron   [pg. 171]

proton, neutron, nucleus   [pg. 165]

proton-proton chain reaction   [pg. 194]

quantum tunneling   [pg. 192]

radioactivity   [pg. 166]

Z: number of protons

Important concepts:

Nucleosynthesis.

Background radiation.

Nuclear chain reaction.

The linear no-threshold model.

Fission and fusion are quantum mechanical processes that can release energy.

Alpha radiation (helium nuclei) is stopped by a sheet of paper.

Beta radiation (electrons) is stopped by a thin sheet of aluminum.

Gamma radiation (high energy photons) is very penetrating.

More than 2000 nuclear bombs have been exploded, the total expenditure on nuclear weapons is estimated at $7.6 trillion.

Irradiating food does not make it radioactive.

FURTHER READING

Richard Rhodes, *The Making of the Atomic Bomb*, Simon & Schuster, 1986.

Robert Serber, *The Los Alamos Primer: The First Lectures on How to Build an Atomic Bomb*, University of California Press, 1992.

EXERCISES

1. Half Life.

   The half life of fermium-252 is one day. If there was 4 kg of fermium-252 30 days ago, how long ago were there 2 kg?

2. Cancer in Colorado.

   Denver receives an annual background radiation dose of 11.8 mSv, which is substantially higher than the US average. Can this be used to test the linear no-threshold model? Discuss complications that may arise.

3. Cancer due to TMI.

   It is estimated that 2 million people received average doses of 14 μSv due to the TMI incident. Evaluate the additional number of cancer cases this caused using the LNT model.

4. Exposure due to Chernobyl

   Compute the dose equivalent to "21 days extra background exposure".

5. Nuclear Weapons.

   (a) Look at Fig. 8.5. How many nuclear bombs does the world need? Justify your argument.

   (b) $7.6 trillion spent on global nuclear weapons sounds like a lot of money. How does this compare to global military expenditure? Global health expenditure? Global research expenditure? What economic impact does such spending have?

   (c) People do not want Iran to build nuclear weapons. Discuss the moral aspects of this desire.

6. LNT.

   It has been argued that using the LNT model is prudent. What do you think of this?

7. Dangerous Radiation Doses.

   The text has quoted two figures: 1 Sv gives a 5.5 % chance of cancer and the odds of dying from a 4.5 Sv dose is 50 %. Are these figures compatible with each other? What else might be going on?

8. CP-1.

   Fermi built the world's first nuclear reactor under an abandoned football stadium at the University of Chicago. Was this irresponsible?

9. Weapons Testing.

   In Sect. 8.5.2 the US government's exposure of millions of people to nuclear radiation from atmospheric testing was deemed reprehensible. Argue *for* this exposure. In retrospect, which argument do you think carries more weight?

10. Radium Quackery.

   Radiactive cure-alls are mentioned in Sect. 8.8. Analyze this phenomenon in light of the discussion in Chap. 3. Why did this brand of quackery fade away?

11. Cats and Dogs.

    Explain why the label "Must not be fed to cats", mentioned on page 190, is especially silly.

12. Kelvin and the Sun.

    Lord Kelvin estimated that the sun could only burn for 3000 years if it were a chemical fire. How would one make this estimate?

A Finite Planet

"Men argue. Nature acts."

— Voltaire

We've put a lot of effort into learning how science works and what it says about a number of issues of concern to many people. In this chapter we will attempt to predict the future characteristics of some of these issues (and a few new ones). It is crucial that we remember that predictions concerning people are only as good as current knowledge and guesses of future behavior.[1] We will start with population growth. It is, of course, people who make demands on the Earth's resources and it is therefore important to know how many people there will be. Even this seemingly simple question is not so simple!

Our next task will be determining how much stuff (water, minerals, fossil fuels) there is and how it gets "used". If we need uranium to supply our energy and uranium is burned within 50 years, it seems important to know this! If we are wise, we will stop burning fossil fuels soon. But even if we follow past behavior and are not wise, fossil fuels will run out eventually. When? What do we do then? Will your grandchildren be driving donkey carts or flying around in personal hover cars?

9.1 Population Growth

I remember when there were 3.5 billion people. Now there are 7 billion. How long can this be maintained? Some people argue that population growth can go on for a long time – we just need to share resources more equitably. This is true, but even good intentions have their limit. We will go back to basics to see just how limited things are.

---

[1] "Prediction is very difficult, especially if it's about the future." — Niels Bohr

© Springer International Publishing Switzerland 2016
E.S. Swanson, *Science and Society*,
DOI 10.1007/978-3-319-21987-5_9

It takes two people to make more people. And the more people there are, the more that can be made. This statement is summarized in a mathematical law that looks like

$$\text{change in population} = r \cdot \text{current population.} \qquad (9.1)$$

The constant r represents the rate at which the population is growing. For example, if every couple had one child per year, and we ignored all issues with age, sex, and consanguinity, then r would be 0.5/yr.

The expression in Eq. 9.1 is called a **differential equation** because it involves a change in a quantity (population in this case). Equations like this can be solved without much trouble. If P is used to represent the population the solution is

$$P(t) = P_0\, e^{r(t-t_0)}, \qquad (9.2)$$

where the initial population is P_0 at time $t = t_0$. This is the famous **exponential law** of population growth.

Notice how different things would be if children were not able to have their own children. In this case (if we imagine people live forever) the only people reproducing would be the original ones and the equation for population growth would be

$$\text{change in population} = R \qquad (9.3)$$

where R is the number of people born per year. The solution to this equation is

$$P(t) = P_0 + R \cdot (t - t_0). \qquad (9.4)$$

This is a **linear equation**, which represents a very different growth law. To see this, let's compare $r = 5\,\%$ exponential growth with linear growth at a rate $R = P_0 r$. The graphs in Fig. 9.1 show both curves running up to $t = 10, t = 100$, and $t = 1000$. Although the growth laws start at the same point and look similar to begin with, exponential growth is absolutely relentless, and soon swamps linear growth. Notice that the scale of the y-axis changes dramatically in each graph.

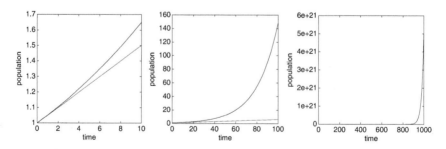

Figure 9.1: Exponential and linear growth.

The notation "2e+21" means $2 \cdot 10^{21}$.

Because exponential growth is so fast, people often reduce it by taking a *logarithm*. This removes the exponential, leaving just the power. For example $\log\exp(\clubsuit) = \clubsuit$ and $\log[P_0\exp(\clubsuit)] = \log(P_0) + \clubsuit$. In this way the thing that is being exponentiated is highlighted, rather than the full exponential growth. A logarithmic plot of the last graph in Fig. 9.1 is shown in Fig. 9.2. Notice that now the exponential growth looks like a straight line, we can actually see the linear growth (which no longer looks straight), and that the units on the y-axis go up by factors of 10^5 for *each* major tic mark.

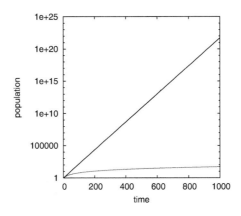

Figure 9.2: Exponential (*solid line*) and linear (*dotted line*) growth on a logarithmic scale.

As a rule of thumb a growth rate of 7 % per year is a doubling every 10 years. Similarly, growth at 3.5 % is doubling every 20 years, and growth at 14 % is a doubling in 5 years.

The world's population is shown on a logarithmic scale in Fig. 9.3. Recall that we expect a straight line if the growth rate is steady. The data are certainly not straight, and in fact, the rate at which the population has been growing has itself been growing! A close look at the figure reveals that something must have happened about the years 5000 BC, 100 BC, and 1950 (look at the inset). We know that it was the Green Revolution that caused the dramatic growth evident in the figure from 1950 to the present. Here we see very plainly the remarkable impact that Norman Borlaug's work had.

Let's think about this population growth. Of course people die as well as are born, so we must refer to a *growth rate* that is given by the birth rate minus the death rate. In 1970 world population was growing at a net rate of 2.2 %. Using Eq. 9.2 and assuming an average weight of 75 kg per person, we can show that the mass of all the people on Earth would equal the mass of the Earth by the year 3070. This ludicrous result demonstrates the incredible power of exponential growth and just how unsustainable it is.

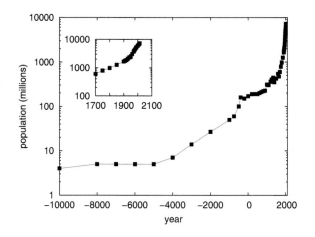

Figure 9.3: World population.

With a 2.2 % growth rate the mass of people equals the mass of the Earth in the year 3070.

Population growth of this sort is very unlikely to happen, in part because the net growth rate is already dropping dramatically. Take a look at Fig. 9.4, which shows something called the *net reproduction rate*. This represents the average number of daughters born to women who reach child-bearing age. A net reproduction rate of one means that each generation will have enough children to replace itself. A rate below one means that the population is declining. The figure shows that North America and Europe have been declining in population for several decades (this does not account for migration), while Asia has just reached break-even. The UN projects that the entire world will approach a net reproduction rate of about one around the year 2160.

> "Democracy cannot survive overpopulation. Human dignity cannot survive overpopulation."
>
> — Isaac Asimov

With these reproduction assumptions in hand, we are ready to make predictions for the world population. The "medium" scenario (green curve in Fig. 9.5) refers to the situation discussed above, where the net reproduction rate stabilizes by the year 2160. The "low" scenario (blue curve) has the population peaking at 7.5 billion people in the year 2040 and then steadily decreasing after that, while the "high" scenario (red curve) postulates continued high birth rates and leads to rapid population growth into tens of billions of people, with an uncertain and tenuous future for all.

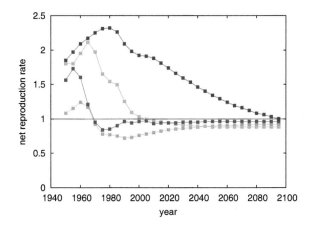

Figure 9.4: Net reproduction rate by region. Africa (*blue*), Asia (*green*), North America (*red*), Western Europe (*orange*). Data past 2005 are 'medium' UN projections.

Source: United Nations, Department of Economic and Social Affairs, Population Division, World Population Prospects: The 2012 Revision, New York, 2013.

9.2 Economic Growth

There is a common assumption that "growth is good" and that things somehow stagnate or even degrade without it. This is most prevalent when talking about economic issues. However, unfettered economic growth leads to ludicrous predictions as surely as unfettered population growth does.

A way to see this is to recognize that it requires energy to do the things that drive economic growth. Thus a growing economy implies a growing (per capita) energy demand. Refer again to Fig. 4.2; the total energy consumed in the USA has been growing exponentially at a rate of about 2.9 % per year since 1700. This is equivalent to a factor of 10 every 79 years. To sustain this growth rate would require covering the *entire planet* with (perfectly efficient) solar panels by the year 2330. Clearly this is not a reasonable expectation!

As if this scenario were not silly enough, all that energy must end up as waste heat (if you have forgotten why, look through Chap. 4 again). This heat will be radiated by the Earth according to the Stefan-Boltzmann law, but not quickly enough to keep the planet from warming. By 2370 the surface would reach the boiling point and there would not be much point discussing whether economies can grow indefinitely or not.

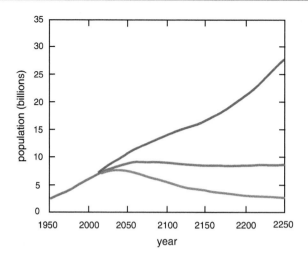

Figure 9.5: Three predictions for world population.

Source: "World Population to 2300", UN Report, 2004.

9.3   Resource Management

"Civilization is in no immediate danger of running out of energy or even ... of oil. But we are running out of environment."

— V. Vaitheeswaran

The problems with population and energy usage growth that we have discussed are rooted in a simple phenomenon: *the Earth is not very big.* Consider that anyone can fly around the world in about 32 hours for a scant $4800.[2]

> The Earth is small.

As usual, when one speaks of sizes in science, one must specify with respect to what. Here, I mean the Earth is small *with respect to us.* Indeed, we have reached the **anthropocene** – the age where the impact of mankind on the planet has become substantial. Biologist E. O. Wilson has calculated that human biomass is one hundred times larger than that of any other large animal species that has ever walked the Earth. And 38 % of the planet's ice-free land is now devoted to agriculture. Fertilizer factories fix more nitrogen from the air, converting it to a

---

[2]I computed this for a Star Alliance trip with itinerary New York, San Francisco, Hong Kong, Dubai, London, New York. The distances of these legs are 4125, 11,100, 5950, 5465, and 5565 km respectively, for a total of 32,205 km.

biologically usable form, than all the plants and microbes on land. Loss of forest habitat is a major cause of extinctions, which are now happening at a rate hundreds or even thousands of times higher than during most of the past half billion years. If current trends continue, the rate may soon be tens of thousands of times higher. And let us not forget our most famous impact: pumping the atmosphere full of carbon dioxide.

The implications of a finite planet will be examined in this section. We will start with an examination of general issues such as how we know how much stuff there is and how one estimates how long it will last. Then we will look at a series of critical resources: water, minerals, and fossil fuels.

9.3.1 Quantifying Resources

Imagine exponential growth that is limited by some physical constraint – a colony of bacteria in a petri dish, for example. In early times the population will grow according to the law of Eq. 9.1. For longer times, however, the constraints start to be felt and hinder growth. A way to model this is to subtract another term that is proportional to the current population squared. Thus one has

$$\text{change in population} = r \cdot \text{current population} - \frac{r}{P_\infty} \cdot (\text{current population})^2 \quad (9.5)$$

Notice that if the population reaches equilibrium, it can no longer change, so the left hand side of this equation is zero. Thus the right hand side tells us that the equilibrium population has to satisfy

$$\text{equilibrium population} = P_\infty.$$

Thus the parameter P_∞ has the simple interpretation of being the long-time equilibrium population.

Once again, this is a simple equation to solve. If the population is denoted P then

$$P(t) = \frac{P_0 \, e^{r(t-t_0)}}{1 + \frac{P_0}{P_\infty}(e^{r(t-t_0)} - 1)}. \quad (9.6)$$

This is called the *logistics function*. Two examples of this function and an exponential function are shown in Fig. 9.6. As promised, for short times the (growing) logistics function looks a lot like an exponential. But the penalty term takes over eventually and forces the population to approach its equilibrium value.

A convenient way to use the logistics function can be obtained from Eq. 9.5. First let us call the current population P and the change in the population $\dot{P}$. Then the equation can be written as

$$\dot{P} = rP \cdot \left(1 - \frac{P}{P_\infty}\right). \quad (9.7)$$

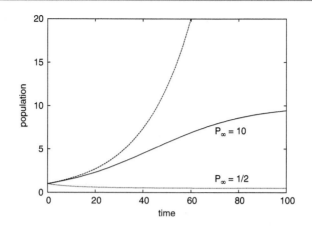

Figure 9.6: The logistics function for $r = 5\%$, $P_0 = 1$, $P_\infty = 10$, or $P_\infty = 1/2$. An exponential function with the same growth rate is shown as a dashed line.

Now divide both sides by P to get

$$\frac{\dot{P}}{P} = r - \frac{r}{P_\infty}P. \tag{9.8}$$

This equation says that if you plot the change in population divided by the current population against the current population then you will get a straight line with y-axis intercept r and an intercept on the x-axis at $P=P_\infty$.

The logistics equation is a model for anything that grows exponentially with a limiting mechanism. Thus it can also describe a *resource*, like oil or uranium, that is being extracted at a rate r, and that has a finite supply, P_∞. In this case P denotes the *extracted resource amount* and $\dot{P}$ is the rate at which it is being extracted. The early exponential growth of the logistics function implies that the resource use grows as the resource becomes more widespread (at the beginning of its exploitation).

Although the logistics model is quite simple, it works surprisingly well for many real resources. For example, the left panel in Fig. 9.7 shows the relative production rate versus production for Pennsylvania anthracite coal. Once production is past 1 Gt the data follow the expected straight line nicely and indicate that the total anthracite available is around 5.7 Gt. The right panel shows total anthracite production by year. The line is the logistics function of Eq. 9.6 (the line fits the data so well that you have to look closely to see it). As can be seen, in spite of intervening wars, strikes, evolving mining technology, and recessions, the logistics model fits the data extremely well with two parameters! I admit to finding this mysterious – sometimes simple models have a power that we do not fully appreciate.

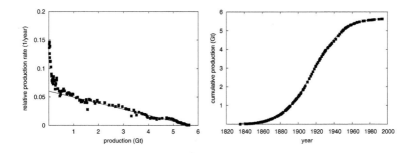

Figure 9.7: *Left*: $\dot{P}/P$ vs. P for Pennsylvania Anthracite. *Right*: Total production with the logistics function.

Source: United States Geological Service.

The logistics model is most readily tested on resources that have run through much of their life cycle, such as Pennsylvania coal. North Sea oil provides another good example, as shown in Fig. 9.8. The graph tells us that there are about 55 Gb of oil in the North Sea and that it will run out in about 2017.

The importance of predictions such as these was illustrated in the recent vote for Scottish independence. At issue was the possible economic viability of an independent Scotland; in particular, YES side politicians predicted that North Sea oil will remain abundant into the far future, while NO side politicians said that it will run out shortly. The public would have been better informed with a graphic such as in Fig. 9.8.

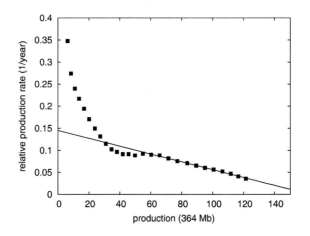

Figure 9.8: $\dot{P}/P$ vs. P for North Sea Oil.

Source: US Energy Information Administration.

Attempting the same analysis with global oil production is less useful because oil production rates have been slowly growing over the past few decades (they are currently about 100 Mb/d). Of course this could be due to economic factors; for example, population growth and increasing prosperity in the third world has led to record prices, and oil producing countries find no reason to not to meet the demand since excess cash flow can be converted into other commodities (like New York real estate or British soccer teams).

But one does not need to rely on production rates to get estimates of reserves. Mining and oil companies actively explore and the rate at which they make discoveries can also be modelled with the logistics function (see Ex. 9). One can argue that discovery rates tend to be more representative of the true resource situation than production rates. We therefore make a logistics plot for crude oil discovery, shown in Fig. 9.9. Again, the model works surprisingly well.

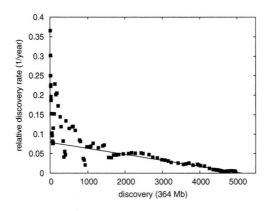

Figure 9.9: $\dot{P}/P$ vs. P for global crude oil discovery.

Source: J. Laherrere ASPO 2005 report with IEA data. [Caveat: I am unable to independently confirm the exploration data.]

If the data represents all oil discoveries (as opposed to reported), then the graph predicts that the world's total recoverable crude oil amounts to 1.9 trillion barrels. Furthermore, the age of discovery of crude oil is essentially over.[3]

With the methodology in place, we are ready to examine global reserves of a number of vital resources.

9.3.2 Water

There are things we use up and there are things we just use. For example, oil, gas, and uranium all get burned and will never be available for our use again. Alternatively, elemental substances such as metals and water remain on the Earth

---

[3]Other types of oil abound, as we shall see.

in one form or another after we are done with them.[4] Thus one does not "waste" water when one runs the tap; what is really being wasted is the effort to transport and clean it.

About 3.5 % of the world's water is fresh, amounting to about $4.8 \cdot 10^{16} \, \text{m}^3$ of water, or 48 million cubic kilometers. About one half of this is tied up in ice sheets, glaciers, and permanent snow. The rest is dominantly ground water. A tiny fraction of all water resides in lakes and rivers. Table 9.1 shows these reservoirs and their "residence times", which is the average time it takes for water to leave the reservoir. Notice that large range of time scales, which is reminiscent of the carbon cycle.

Table 9.1: Reservoirs of water and their residence times.

Body	Volume (km^3)	Residence time
Oceans	1,338,000,000	3,200 years
Ice and snow	24,064,000	20,000 years (Antarctica)
		20–100 years (glaciers)
Fresh ground water	10,530,000	100–200 years (shallow)
		10,000 years (deep)
Saline ground water	12,870,000	"
Soil moisture	16,500	1–2 months
Permafrost	300,000	–
Fresh water lakes	91,000	1–100 years
Saline lakes	85,400	50–100 years
Atmosphere	12,900	9 days
Swamps	11,470	–
Rivers	2120	2–6 months

Source: USGS

The central issue with water as a resource is *availability*. About 70 % of withdrawn water is used for agriculture, industry uses 20 %, and 10 % is personal.[5] Each of these uses presents its own problems. Water used for industry and in the home tends to be returned to its source very quickly; but it is (often) returned dirty. It is only in the past 50 years that we have become aware of this problem. Water polluted with sewage has changed coastal regions and bays in much of the world. Many cities in the United States still struggle with separating runoff water from sewage. Industrial waste water used to be dumped in rivers and lakes as if these were infinitely large reservoirs. Eventually pollution by heavy metals, nitrates and phosphates, and dozens of other noxious materials turned lakes and rivers into little more than sewers.

---

[4]Within limits. It is possible for molecules such as water to leave Earth. This happened on Venus and Mars.

[5]Source: UN World Water Development Report 3, *Water in a Changing World*, 2009.

Ex. The recent history of Lake Erie provides a prime example of the damage that can be done. The lake became very polluted in the 1970s as a result of industrial waste and sewage runoff. Every day Detroit, Cleveland, and other towns dumped 1.5 billion gallons of inadequately treated waste into the lake. Fertilizer from farming contributed large quantities of nitrogen and phosphorus, which causes *eutrophication*. This a bloom in algae and other plants due to excess nutrients. Decomposition of this plant life kills fish by depriving them of oxygen. As a result fish kills and fecal matter frequently fouled the shores of the lake. Perhaps the most infamous occurrence during this period was when Cleveland's Cuyahoga River caught fire, indicating just how sludgy the river had become (in fact, the river had caught fire at least ten times since 1868).

This and other embarrassments finally forced politicians to react. The Environmental Protection Agency was created by President Nixon in 1970 and the Clean Water Act was passed in 1972. Substantial effort (see Table 9.2) has since been expended or is planned to remediate past sins. These efforts are paying off, and Lake Erie, for one, now supports wildlife again.

Table 9.2: Committed water remediation funds.

Body	Cost ($ billions)
Everglades	10.9
Upper Mississippi	5.3
Coastal Louisiana	14
Chesapeake Bay	19
Great Lakes	8
California Bay Delta	8.5

UN World Water Development Report 3, *Water in a Changing World*, 2009.

The 70 % of water that is used for agriculture creates its own problems. Chief amongst these is that this water is removed from the surface water reservoir and either enters the atmosphere through evaporation or enters the food chain. Thus this water is no longer available for immediate use (although it mercifully remains clean). Increasing population has led to increasing cultivation of marginal lands that require extensive irrigation. Chronic water shortages in many regions of the world are the result. Figure 9.10 shows the Colorado River as it passes the US-Mexico border. The Canal Alimentador Central runs off to the West. The trickle that continues is the remnant of the river. Even this does not gain the dignity of making it the remaining 60 miles to the Sea of Cortez.

Lack of surface water leads many places to pump water from underground *aquifers*. In fact 20 % of global water use is from aquifers. Often these aquifers are replenished by surface water seeping through the soil, so this is a sustainable

Figure 9.10: The Colorado River.

Google 2014, Cnes/Spot Image, DigitalGlobe, Landsat, US Geological Survey, USDA Farm Service Agency. Map data: 2014 Google, INEGI.

activity if done slowly enough. Unfortunately, many aquifers are *fossil water*, meaning that they were created long ago and have no ability to be replenished. Once fossil water aquifers are empty other sources of water must be found. A recent study[6] has found that 80 % of the world's aquifers are being used sustainably but this is offset by heavy over-exploitation in a few key areas. Those areas include western Mexico, the American Plains, California's Central Valley, Saudi Arabia, Iran, northern India, and parts of northern China. Altogether about 1.7 billion people live in areas where groundwater use is greater than replenishment rates.

> Ex. Libya's Great Manmade River project was built at a cost of $25 billion to supply Tripoli, Benghazi, and Sirte with an estimated 6.5 Mm^3 of water a day. The network of pipes and boreholes is sucking water out of the ground that was deposited in the rocks under the Sahara 40,000 years ago, but is not being replenished. It is estimated the aquifer will run dry in 60–100 years.

To be clear, there is no shortage of fresh water – only an imbalance of supply. Approximately 107,000 km^3 of precipitation falls on land every year (this is about 0.2 % of the globe's fresh water). Of this, about 30 % ends up in rivers. Thus rivers can supply about 4000 m^3 of water per person per year. This is substantially more than the global average water use for all purposes of about 1200 m^3 per person per year (estimate from the UN World Water Development Report).

Unfortunately, water supply imbalance is a serious problem. More than 60 % of the world's population growth between 2008 and 2100 will be in sub-Saharan

---

[6]T. Gleeson, *et al.* *Water balance of global aquifers revealed by groundwater footprint*, Nature, **488**, 197 (2012).

Africa (32 %) and South Asia (30 %). Together, these regions are expected to account for half of world population in 2100. About 340 million Africans lack access to safe drinking water, and almost 500 million lack access to adequate sanitation. Countries in sub-Saharan Africa store only about 4 % of their annual renewable flows, compared with 70–90 % in many developed countries, yet water storage is essential to ensure reliable sources of water for irrigation and drinking. This lack of infrastructure is tied to the deaths of 1.4 million children per year from diarrheal diseases. For some people water shortages mean more than a dry lawn.

9.3.3   Minerals

Maybe you have heard that we are "running out" of gold or aluminum. As with water, minerals never leave the environment, they just change form from useful to "less useful" (i.e., rusted, less dense, contaminated, etc.).[7] Thus there is a sort of "second law of mineral use",

> The usefulness of a manufactured object always degrades with time.

Of course, this situation can be reversed with the judicious application of energy to recycle "used" materials.

We seek to estimate global resources of various quantities in this section. Thus, if there are 7 billion ounces of gold in the world, everyone's allotment is one ounce. If the population doubles, this allotment halves. Unfortunately, little effort appears to have been put into estimating total global resources.[8] Instead, companies and government agencies tend to take the direct approach of adding up what people think is in the ground. Thus resources are often reported as *reserves*, which is a company's known working inventory. In 1970 world copper reserves were 280 million tonnes, while the estimated resource was 1.6 billion tonnes.

Estimating reserves and resources is complicated by economic factors. For example, bismuth is the 69th most common element on Earth at around 8 ppb in the crust. But it tends to be well-mixed and is therefore difficult to extract. Resources that are too difficult or expensive to extract are rarely counted in reserves. In this sense, there will always be gold available, it will just be extremely difficult to obtain. At that stage it is of course cheaper to recycle previously mined materials.

Every year the US Geological Survey collects information from the world's mining companies on production and reserve estimates. I have used their data to compile Table 9.3. Vital energy-related resources will be discussed separately

---

[7]Except helium, which floats to the top of the atmosphere and is permanently driven away by "solar winds".

[8]As we have seen, this is possible if discovery rates are tracked, but I can find no such data.

later on. The last two columns are the ratios of the resource amount and reserves to the current production rate. These are rough estimates of when mining operations will stop and recycling operations must begin. Production rates are rarely stable, and one must expect a period of accelerating production followed by declining production (such as is accounted for by the logistics model), thus these estimates should always be treated as approximate.

Table 9.3: Global mineral resources, 2013.

Resource	Amount (Mt)	Reserves (Mt)	Production 2013 (kt/yr)	A/P (years)	R/P (years)
Antimony		1.8	163		11
Gold		0.054	2.77		19
Silver		0.52	26		20
Tin		4.7	230		20
Nickel	130	74	2490	52	30
Bismuth		0.32	7.6		42
Tungsten		3.5	71		49
Mercury	0.6	0.094	1.81	331	52[e]
Niobium		>4.3	51		>84
Titanium[d]	>2000	700	6790	>295	103
Copper	3500		20,000		175
Vanadium	>63	14	76	829	184
Palladium	0.15[a]	0.066[a]	211	<474	<313
Platinum	0.15[a]	0.066[a]	192	<521	<344
Lithium	73.8	13	35	2100	371
Rare earths[f]		140	110		1270
Aluminum			47,300		b
Indium		c	0.77		
Phosphate	>300,000	63,000	2240	>134k	30k

Source: US Geological Service.
Notes:

(a) Estimate for platinum group metals.

(b) "World reserves for bauxite are sufficient to meet world demand for metal well into the future."

(c) "Quantitative estimates of reserves are not available."

(d) Ilmenite.

(e) "The declining consumption of mercury, except for small-scale gold mining, indicates that these resources are sufficient for another century or more of use."

(f) Rare earths are elements such as scandium, yttrium, and the lanthanides. China controls 90 % of production and has 40 % of global reserves.

As mentioned, helium is a special case because it is lost permanently when it is "used". Current use is 171 million m^3/yr. Helium is chiefly produced as a by-product in natural gas extraction. It is estimated that global resources are 57.9 billion m^3, which gives a ratio $A/P = 338$ years.

Antimony is the most imperiled of the metals. Its most important use is as a hardener in lead batteries. If we decide that antimony is important, the industry producing it must change from mining to recycling. Because of this it is essential that antimony-containing products be disposed of in a way that permits easy future retrieval of the substance. Naturally, the same can be said of all the other materials in this table.

9.3.4   Fossil Fuels and Energy

You will recall (Chap. 7) that we burn carbon at a prodigious rate and that we like doing it because it raises our effective wealth. As of 2012 the average Briton consumed about 16 kg of fossil fuels per day, creating about 11 tonnes of waste carbon dioxide per year. How much longer will the boom times last? And just how much CO_2 will end up in the atmosphere?

To answer these questions we turn to the *2010 Survey of Energy Resources*, produced by the World Energy Council. In their own words the WEC is, "the UN-accredited global energy body, representing the entire energy spectrum, with more than 3000 member organisations located in over 90 countries and drawn from governments, private and state corporations, academia, NGOs and energy-related stakeholders." Fair enough.

Table 9.4 summarizes the WEC estimates of global fossil fuel resources. These are "recoverable" resources, namely they are economically feasible to remove. Given the thoroughness of global oil exploration, one can think of these figures as describing something between "resource" and "reserve" amounts.

Table 9.4: WEC global fossil fuel resources.

Resource	Amount (Gt)	Production (Gt/yr)	A/P (years)
Natural gas	132	3.0	44
Coal	861	7.83	110
Crude oil & NGL	191	5.0	38
Shale oil	738	a	
Bitumen & heavy oil	902	a	328

Notes: I have assumed 6.5 barrels crude oil = 1 t; 6.1 barrels heavy oil = 1 t; 1 m^3 natural gas = 0.714 kg, mostly from CH_4. The WEC coal estimate is lower than that of the German Federal Institute for Geosciences and Natural Resources, which estimates 1038 Gt.

(a) I assume that shale oil and heavy oil will supplant crude in future and therefore take the current rate of use of these as 5 Gt/yr. This rate will be higher once gas and coal run out.

It is clear that current practices will end in a century or less. However, unconventional fuel sources (shale oil and bitumen) promise hundreds of years of continued fouling of the atmosphere. Is it time we shook the fossil fuel habit?

9.3.5   Fossil Fuels and Temperature

As we have stressed in Chap. 7, burning fossil fuel is not cost-free. In particular, burning 1 kg of coal creates 2.9 kg of carbon dioxide, of which about 1.0 kg is retained in the atmosphere.

> Ex. Let's check these numbers for oil. Oil (and coal) are about 78% carbon. We have said previously that there are 6.5 barrels of crude oil per tonne; thus burning a barrel of oil releases $1/6.5 \cdot 0.78 \cdot 44/12 = 0.43$ tCO$_2$/barrel. See Ex. 15 to supply the missing units.

In Chap. 7 we examined the IPCC's worst case "drill baby drill" RCP8.5 scenario. Frankly I do not find this scenario pessimistic enough. I therefore wish to examine the "judging by past behavior" scenario in which every shred of cheap carbon is burned. Total resources are presented in Table 9.4. As we have seen, each burnt kilogram releases about 0.8 kg of carbon into the environment. The resulting increase in average temperature can be obtained from a figure in the IPCC AR5 report that shows the dependence of temperature on CO$_2$ emissions (Fig. 9.11). The extrapolation in the JPB scenario arrives at a final temperature anomaly of 6 C (= 10.8 F) and is indicated as a star in the figure.

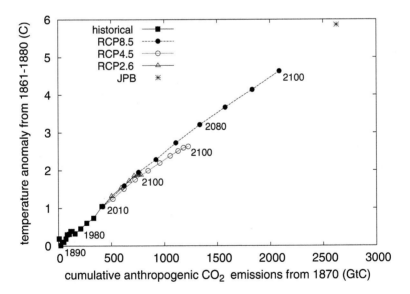

Figure 9.11: Temperature anomaly vs. total added CO$_2$.

Source: IPCC AR5.

To understand what this means, consider the following map of average US temperatures (Fig. 9.12). An increase of 10 degrees F shifts the zones two places.

Thus the typical weather in Delaware will be like the current weather in Jacksonville, FL. While we are at it, the IPCC states that it is very likely that the Greenland ice sheet will melt in this scenario, which will cause the sea level to rise 7 m (23 ft).

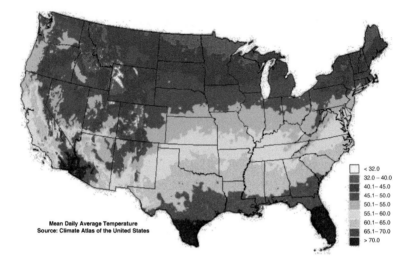

Figure 9.12: US mean daily temperatures (1961–1990).

9.3.6   Fossil Fuels and Air Pollution

Between October 27 and October 31 1948 the mill town of Donora, PA was enveloped in a large temperature inversion. The resulting smog killed 20 people and sickened one third of the town. The subsequent publicity raised public awareness of the issue of air pollution and, along with various disasters in US waterways (Sect. 9.3.2), helped forge the way to the foundation of the Environmental Protection Agency.

Coal burning power plants are a major contributor to the nation's smog load. Estimates for various noxious products due to coal combustion are listed in Table 9.5. Of course, burning other fuels, especially diesel, also creates pollutants.

Since the 1970s studies have revealed that smog can have serious negative health effects. Of concern is ***particulate matter*** that remains airborne as aerosols for lengthy periods. Particles smaller than $10 \, \mu m$ are especially dangerous because they are easily inhaled and can reach alveoli. Particles smaller than $2.5 \, \mu m$ can penetrate into the gas exchange regions of the lungs and cause DNA mutations, heart attacks, hardening of the arteries, and lung cancer. These particles

Table 9.5: Pollutants emitted by US coal plants in 2011.

Pollutant	Amount	Amount with scrubbing
Sulfur dioxide	6.8 Mt	3.4 Mt
Nitrogen oxides	5.0 Mt	1.6 Mt
Particulate matter	0.2 Mt	2400 t
Mercury	37 kt	3.7 kt
Lead	25 kt	2.5 kt
Arsenic	50 kt	–

Source: Union of Concerned Scientists.

are called PM10 and PM2.5 respectively. It is estimated that particulate matter pollution causes 22,000–52,000 deaths per year in the United States.[9]

As a result, the EPA has set safe levels of PM2.5 at $15 \, \mu g/m^3$. Figure 9.13 shows the average density of PM2.5 between 2001 and 2006 as determined from satellite data. Dark areas in the east exceed the EPA limit; as can be seen, these tend to concentrate on large urban areas, the central valley of California, and a belt reaching from Ohio, through Indiana, and along the Mississippi valley. Globally, regions of high PM2.5 concentration are the desert regions of north Africa and the Middle East, northern India, Pakistan, and central Europe. Northern China is the world's smoggiest place with PM2.5 concentrations regularly exceeding $80 \, \mu g/m^3$. We note that China accounts for 19 % of world population but 27 % of world cancer.

The effect of particulate matter on public health can be examined with cohort studies (Sect. 2.3.1). Examples of typical results are

> Ex. "The research is based on long-term data compiled for the first time, and projects that the 500 million Chinese who live north of the Huai River are set to lose an aggregate 2.5 billion years of life expectancy due to the extensive use of coal to power boilers for heating throughout the region."
>
> "Every additional 100 micrograms of particulate matter per cubic meter in the atmosphere lowers life expectancy at birth by three years."
>
> Y. Chen *et al.*, *Evidence on the impact of sustained exposure to air pollution on life expectancy from China's Huai River policy*, PNAS, **110**, 12936 (2013).

> Ex. "The gap in life expectancy between areas with good air quality and moderately heavily polluted areas was 3.78 years for women of age 65 and 0.93 years for men."
>
> M. Wen and D. Gu, *Air pollution shortens life expectancy and health expectancy for older adults: the case of China*, J. Gerontology, Biol. Sci., and Med. Sci., **67**, 1219 (2012).

---

[9] A.H. Mokdad *et al.*, *Actual Causes of Death in the United States, 2000*, JAMA, **291**, 1238 (2004).

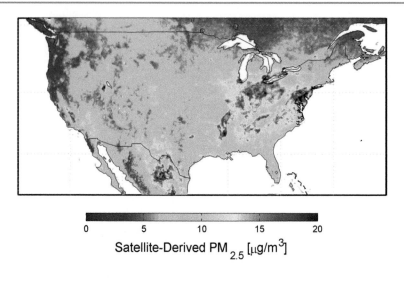

Source: van Donkelaar A., Martin R.V., Brauer M., Kahn R., Levy R., Verduzco C. *et al.*, "Global estimates of ambient fine particulate matter concentrations from satellite-based aerosol optical depth: Development and application", Environ. Health Perspect. **118**, 847 (2010). Reproduced with permission.

Ex. "A decrease of 10 μg per cubic meter in the concentration of fine particulate matter was associated with an estimated increase in mean life expectancy of 0.61 ± 0.20 year (P=0.004)."

C.A. Pope, M. Ezzati, and D.W. Docke, *Fine-Particulate Air Pollution and Life Expectancy in the United States*, New England Journal of Medicine, **360**, 376 (2009).

Data from this last paper are presented in Fig. 9.14. The letters refer to metropolitan areas in the USA. The difference between Steubenville OH (STE) and Denver CO (DEN) is $20 \mu g/m^3$. The corresponding life expectancies are 73.5 and 75.2 years. These data are for the years 1979–1983; data from 2000 indicate much lower PM2.5 concentrations and much longer life expectancies. Fits to the data (dotted line) gave a reduction in life expectancy of 1.19 ± 0.27 years per $10 \mu g/m^3$ PM2.5 in 1982 and 2.02 ± 0.50 years per $10 \mu g/m^3$ in 2001. Of course these raw results (i) do not indicate causality, (ii) do not account for confounding variables (see Ex. 18). Taking confounding variables into account yielded the result quoted in the example.

Life expectancy is reduced 7 months for every $10 \mu g/m^3$ of particulate matter.

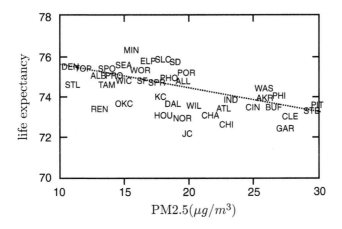

Figure 9.14: Life Expectancy vs. particulate pollution for 1975–1983.

Source: C.A. Pope, M. Ezzati, and D.W. Docke, *Fine-Particulate Air Pollution and Life Expectancy in the United States*, New England Journal of Medicine, **360**, 376 (2009).

9.4 Beyond Carbon

It would be foolish to agitate for moving beyond the fossil fuel era without a rational assessment of alternative sources of energy. In this section we examine the prospects for hydroelectric, wind, solar, nuclear, and fusion power. A major part of viable future plans must include weaning the transportation sector from its addiction to carbon. This is discussed in Sect. 9.5.4.[10]

To set the scale, the current (2008) global energy demand is 144,000 TWh/yr. In the units we have been using this is 16 TW. This amounts to 2.3 kW/person. The per capita energy use has been growing at about 0.5 %/yr over the past 20 years. Over this same time the population has grown at 1.3 %/yr so that total energy growth is around 1.8 %/yr.

World energy demand is 16 TW = 2.3 kW/person, growing at 0.5 %/yr.

9.4.1 Hydro

Hydroelectric power is a relatively benign source of power – the chief environmental cost is the loss of land for the reservoir and possible issues with upstream silting and downstream habitat change.[11]

---

[10]In my opinion biofuels and hydrogen cars are silly. We will not discuss them.

[11]And the salmon. We can't forget the salmon.

The World Energy Council (Sect. 9.3.4) states that hydropower is being utilized in more than 160 countries with a net capacity of 874 GW in 2008 (125 W/person). It is estimated that 1.8–2.6 TW (260–370 W/person) of economically feasible hydropower is available world wide. The largest potential gain can be realized in Africa, which has only exploited 10 % of its hydropower capability.

> One can calculate the theoretical maximum hydropower available in a region by taking its average rainfall and multiplying by the acceleration due to gravity and the average height of the land. Of course, the result will be much larger than is typically practical.

9.4.2   Wind Power

Wind provides another relatively benign source of power. In this case the environmental costs are an unsightly skyline, possible local noise, and bird kills.[12] By the way, the things that make power from wind are called **wind turbines**, not wind mills, which are wind-powered grain grinders.

Current global capacity is 318 GW (45 W/person). As Fig. 9.15 demonstrates, this capacity is growing rapidly. If this growth continues one can expect to hit 16 TW in the year 2032.

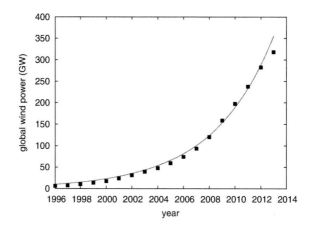

Figure 9.15: Cumulative global wind power. The dotted line is growth at 21 %/yr.

Source: Global Wind Energy Council, *Global Wind Statistics, 2012.*

It is possible to estimate the area required for such prodigious power with some simple arithmetic. A wind turbine extracts kinetic energy from the air by turning blades, which turn the armature of a generator, which makes AC current

---

[12]Birds killed by wind turbines are a fraction of those killed by cars, which is a fraction of those killed by domestic cats.

(see Sect. 5.3). If one imagines that air at speed v comes into a turbine with area A and leaves with zero speed, then the maximum energy available to generate electricity is the kinetic energy of a volume of air of area A and length vt, where t is the time we are considering. Thus the available energy is

$$\frac{1}{2}m_{air}v^2 = \frac{1}{2}\rho A \cdot vt \cdot v^2 \tag{9.9}$$

where ρ is the density of the air. Divide this by t to obtain the power. Since typical wind turbine efficiency is $\epsilon = 50\%$, the power per turbine is

$$\frac{1}{2}\epsilon\rho Av^3. \tag{9.10}$$

Experts say that the closest turbines can be placed to each other is around $(10r)^2$, where r is the length of a turbine blade. Thus the power per wind turbine per area required is

$$\frac{\frac{1}{2}\epsilon\rho(\pi r^2)v^3}{(10r)^2} = \frac{1}{200}\epsilon\pi\rho v^3 = 0.01(v \text{ m/s})^3 \text{ W/m}^2. \tag{9.11}$$

I've used $\rho = 1.3\,\text{kg/m}^3$ as the density of air and taken $\epsilon = 0.5$ to obtain the last equation.

Average wind speed varies with height. At $10\,\text{m}$ it is $6.64\,\text{m/s}$ over the oceans and $3.28\,\text{m/s}$ over land. At $80\,\text{m}$ these are 8.60 and $4.54\,\text{m/s}$ respectively.[13] Assuming an average wind speed of $4.5\,\text{m/s}$ gives[14] a maximum power density of $0.9\,\text{W/m}^2$. Generating $16\,\text{TW}$ of power would then require $1.8 \cdot 10^{13}\,\text{m}^2$ of land. This should be compared to the world's land area of $1.5 \cdot 10^{14}\,\text{m}^2$.

Although present growth rates appear to point to wind energy as a solution to our energy needs, covering 12% of the world with wind turbines does not seem feasible. One could consider placing turbines at sea, but this raises costs. Wind, it seems, can only take us part of the way to a carbonless future.

9.4.3 Solar Power

A number of technologies that capture energy from sunshine exist. Of course plants are the original technology. While there are several facilities that use large arrays of mirrors to focus light on boilers in the US, Australia, and Spain we shall concentrate attention on ***photovoltaic technology***.

The photovoltaic effect is similar to the photoelectric effect (Sect. 6.4) in that incident light is capable of freeing electrons from the medium. In this case the

---

[13]C.L. Archer and M.Z. Jacobson, *Evaluation of Global Wind Power*, J. Geo. Res., **110**, D12110 (2005).

[14]This is not quite the correct procedure. See if you can figure out why.

electrons join the **conduction band** of the material, which permits them to flow over large distances as electricity. The photovoltaic effect was discovered by Edmond Becquerel (1820–1891), the father of Antoine Bequerel, in 1839.

There are currently (2013) 139 GW (20 W/person) of installed solar energy generation facilities world wide. This figure has been growing at an astonishing 48 %/yr recently, in part due to government incentive programs.

Figure 7.6 indicates that incoming solar energy density is around 340 W/m^2 and typical available energy at ground level is 186 W/m^2 (this is the amount absorbed and reflected by the surface). This figure will change depending on where you are, as shown in Table 9.6.

Table 9.6: Average insolation by locale.

Locale	Average insolation (W/m^2)
Anchorage, AK	87
Oslo, NO	94
London, UK	109
Paris, FR	125
New York, NY	147
Pittsburgh, PA	147
Boston, MA	149
Chicago, IL	155
Atlanta, GA	182
Salt Lake City, UT	189
Houston, TX	197
San Francisco, CA	204
Miami, FL	219
Los Vegas, NV	221
Phoenix, AZ	224
Cairo, EG	237
Nouackchott, MR	273

Source: D. MacKay, *Sustainable Energy*, UIT Cambridge, 2009.

Modern photovoltaic (PV) panels operate at about 10 % efficiency, while more expensive panels can reach 20 %. Let us therefore assume a global average power density of 30 W/m^2. To achieve 16 TW would thus require $5.3 \cdot 10^{11}$ m^2 of land, or about 0.35 % of all land must be covered in PV panels. Alternatively, if solar facilities are built in the Sahara and efficiencies reach 20 %, this figure drops to 0.19 % of all land. This is equivalent to a square of size 540 × 540 km.

At an installed cost of \$4.00/W, this project will cost 64 trillion dollars. Unfortunately, desert regions tend to be where people do not live, and the produced energy will need to be distributed to populated areas. This will require the construction of another enormous and expensive infrastructure.

9.4.4 Nuclear Power

As we have seen (Sect. 8.6), nuclear reactors require natural uranium-238 or enriched uranium-235. Since the uranium is burned (via Eq. 8.7), it is a stretch to label nuclear power "renewable" or "sustainable".[15] However, the waste produced by nuclear power is self-contained and of a readily manageable volume (Sect. 8.6.3). Let us therefore examine future prospects for nuclear power.

Current nuclear power capacity from the world's 434 reactors is 373 GW (53 W/person). This is set to grow by 77 GW as 73 reactors are under construction world wide. These reactors consumed 65.9 kt of uranium per year in 2014. Table 9.7 shows global resources based on extraction cost and the degree of speculation used to form the estimate.

Table 9.7: Global uranium reserves, 2009.

Quality	<$130/kgU (MtU)	<$260/kgU (MtU)	R/P ($260/kgU) (years)
Assured	3.53	4.00	61
Inferred	1.88	2.30	35
Speculative	6.55	6.80	103
Unconventional	–	10–22	152–334
Ocean	–	4000	60 k

World Energy Council, *2010 Survey of Energy Resources* and Nuclear Energy Institute figures.

These R/P estimates are for once-through light water reactors that treat burned fuel as waste. However, recall (Sect. 8.6.2) that spent fuel bundles are far from useless, and are recycled in France and other countries. In the same section we also discussed breeder reactors that are capable of generating new fuel. Thus, although 96 years of nonspeculative reserves looks alarming, this span can be stretched considerably (possibly up to 1000 years if breeder reactors are used exclusively starting now). Furthermore, it appears that another 100–400 years of light water reactor operations are possible with more speculative sources. Lastly, uranium exists at 3 ppb in sea water. If it is possible to extract this in a reasonably energy-efficient manner, then it will be possible to generate nuclear power well into the future.

If you are inclined to think that nuclear energy solves global power problems, recall that we need a capacity of 16 TW (and growing). This power outlay requires 43 times as much uranium per year as is currently consumed. In this scenario all uranium except that in the oceans would be consumed in 12 years. It therefore appears shortsighted to treat spent fuel as waste.

---

[15]Come to think of it, solar power is not sustainable either! See Sect. 8.9.3.

> Spent nuclear fuel should be stored for future generations.

We have discussed how thorium can be used as an alternate fuel in Sect. 8.6.2. The waste product of thorium fuel cycle is uranium-233, which is also fissile. Thus thorium reactors can be run as breeder reactors. The World Energy Council estimates a reserve at 6 MtTh, but this is likely quite conservative as thorium is thought to be three times more abundant than uranium in the Earth's crust. Thus thorium and thorium reactors are important possible sources of future electricity.

9.4.5 Fusion Power

We have seen how the fusion of light nuclei can release energy, and that it is this energy that powers the sun (Sect. 8.9.3). Even before the details were worked out, Sir Arthur Eddington speculated about releasing this energy for man's use on Earth (pg. 191).

Research on fusion reactions started during the Second World War, spurred by both the desire to generate energy and to build bombs. Uncontrolled fusion was demonstrated with the explosion of the hydrogen bomb called Ivy Mike in 1952. Meanwhile, research on energy generation continued outside of the military at Princeton University, at Los Alamos National Laboratory, at the Atomic Energy Research Establishment in Harwell, UK, and in the Soviet Union.

The simplest fusion reaction being considered is called the **DT reaction** because deuterium and tritium act as the fuel. The reaction is

$$\,^2_1\mathrm{H} + \,^3_1\mathrm{H} \rightarrow \,^4_2\mathrm{He} + \mathrm{n} + 17\,\mathrm{MeV}. \tag{9.12}$$

The reaction will not proceed unless the fuel is at high temperature and density (simulating the core of the sun). Achieving this has proved a challenging technical obstacle, and the field has experienced slow progress over the past 60 years.

Current effort is centered at the National Ignition Facility at Lawrence Livermore National Laboratory in California and at the ITER facility in Cadarache, France. The NIF seeks to confine and heat pellets of deuterium and tritium by blasting them with 1.8 MJ of energy from 192 laser beams with the hope that a self-sustaining fusion reaction will occur. On September 28, 2013 the NIF succeeded in causing a fuel pellet to give off more energy than was immediately applied to it. Unfortunately this was 14 kJ, significantly less than the 1.8 MJ that was required to initiate the reaction.

The ITER facility is a huge multinational program to build a magnetic confinement fusion reactor called a **tokamak**. The idea is to confine a hot plasma with magnetic fields, which are required since no material can withstand the heat of the plasma. The reactor has a design goal of producing 500 MW of power. Construction will be done in 2019, experiments will start in 2020, and it is hoped that DT fusion will be achieved by 2027. The goal of generating electricity remains in the distant future since ITER is an experimental machine.

Fusion reactors hold great promise for future energy generation. Their fuel is readily obtainable and radioactive waste is negligible. A 1 GW fusion plant will require about 100 kg of deuterium and 3 tonnes of lithium to operate for a year. There is also no danger of runaway disasters since any deviations from running conditions simply shut down the reaction. The biggest potential issue is the presence of tritium, which has a half life of 12 years and can easily escape to the environment. Avoiding tritium entirely is possible with a DD fusion reaction, but this requires even higher temperatures than the DT reaction. Finally, plasmas are so hot that equipment malfunction that permits plasma escape could destroy the reactor. This would be an economic disaster for the agency managing the facility, but would not be an environmental disaster like Chernobyl.

9.5 Electric Transport

If we are lucky, a combination of wind, solar, and nuclear power will enable us to squelch our fossil fuel addiction. But our hopes have a gaping hole: transportation comprises 26 % of energy use (Fig. 4.3), and this relies wholly on fossil fuels. A lot of silly things have been said about this – maybe we can grow biofuels, or use hydrogen – but making these fuels consumes vast amounts of energy. Why not use that energy to drive vehicles directly?

9.5.1 The Public Transit Option

The only sensible plan is to electrify the national (global!) transportation system. A simple way to achieve this is to ban most vehicles and rely on electric trams and subways for commuting and electric trains for long distance travel. This would have the great benefit of allowing centralized electricity production and would introduce a dramatic reduction in energy requirements because public transit is far more efficient than personal transit.

Let's see what is involved. From Fig. 4.3 we learn that 24.7 quads of energy are used per year to power 5.6 quads of transportation. This corresponds to an efficiency of 5.6/24.7 = 23 %, which is a reasonable estimate of the efficiency of a typical internal combustion engine. About one half of the wasted energy is emitted as exhaust heat; the other half goes into heating other components of the car.

There is good news for our electrification scheme: electric motors are far more efficient than internal combustion engines! Efficiencies between 80 % and 90 % are possible. If we use an efficiency of 85 % we require 5.6/0.85 quads/yr = 0.21 TW of power to run the nation's transportation system. There is another savings because public transit is more efficient than personal transit. We will quantify this by specifying the energy required to move one person one km. The units are then J/p · km. If a car uses gas at 30 miles per gallon and carries four people its rating would be

$$\frac{1}{30}\frac{\text{gal}}{\text{mile}} \cdot 0.61\frac{\text{miles}}{\text{km}} \cdot 2.76\frac{\text{kg}}{\text{gal}} \cdot 42\frac{\text{MJ}}{\text{kg}} = 2.36\frac{\text{MJ}}{\text{km}}. \qquad (9.13)$$

(See Ex. 1 of Chap. 7 or Table 9.9 for the energy density of gasoline.) This works out to 0.6 MJ/km per person. Figures for typical buses and trains are given in Table 9.8.

Table 9.8: Energy requires for different modes of transit.

Transport	Energy consumption (MJ/p· km)
Car	0.59
Bus	0.22
High speed train	0.11

Source: D. MacKay, *Sustainable Energy*.

These figures are unrealistic because they assume full vehicles at all times. Data collected from Japan indicate that practical values are 2.4 MJ/p · km (cars), 0.7 MJ/p · km (bus), and 0.22 MJ/p · km (rail). If we can get people to take public transit one half of the time, the national budget for transportation energy will drop from 0.21 to 0.13 TW. All told we have reduced power requirements in this sector from 0.91 to 0.13 TW and will have eliminated fossil fuel contributions to air pollution and halted global warming.

The problem with this dream is that it does not match well with the American lifestyle. Personal mobility is something people treasure; even Europeans – who rely on mass transit far more than North Americans – love their cars. Let us therefore examine the only sensible carbon-free option: the personal electric vehicle.

9.5.2   Batteries

Electric cars have a luxurious feel to them: they accelerate smoothly and powerfully, are whisper-quiet, don't smell, and don't pollute. Unfortunately, they are powered by batteries, and batteries are ugly. More specifically, they have a very low ability to carry energy compared to gasoline. The *specific energy* and *specific power* for several types of battery are given in Table 9.9. The factor of 100 or 1000 that must be made up in specific energy over gasoline is a serious impediment to our electrification aspirations. In particular, vehicle range becomes an issue. To make matters worse, batteries take a long time to recharge (fractions of a day if special facilities are not available). About the only good news is that their specific power is comparable to gas, and electric vehicle acceleration is therefore similar to that of carbon belchers.

Table 9.9: Characteristics of various batteries.

Type	Specific energy (kJ/kg)	Specific power (W/kg)
Lead acid	146	20
Alkaline	400	5–100
Carbon-zinc	130	3–24
NiMH	250	150–400
NiCad	140	4–200
Lithium-ion	300–720	300–1500
Butter	37,600	–
Gasoline	42,000	150–400

Batteries are complex electro-chemical devices. Their energy and power output depend on temperature, current, and even past charging patterns. Thus the figures in Table 9.9 are approximate and are displayed mainly to illustrate the point that batteries are power-rich and energy-poor.

If you look at your cellphone battery you will see a notation like "2460 mAh". This is a current capability measure, which is similar to the energy content of the battery. It is often interpreted as meaning the battery is capable of supplying 2.46 amps of current for an hour, or 1.23 amps for 2 h, etc. In reality the charge (or energy[16]) that can be supplied by a battery depends on the rate at which it is drawn (amps, or watts if we think in terms of energy). The available energy vs. draw rate for an alkaline and a lithium battery are shown in Fig. 9.16 on a log-log scale. Notice that energy is specified in $W \cdot h$, which is $1 J/s \cdot 3600 s = 3.6 kJ$. The curve indicates that batteries supply a uniform amount of energy for any draw rate, but that this supply degrades rapidly past some point. This nonlinear behavior has practical implications. For example, alkaline batteries are best used in devices, such as flashlights, that have a steady and low power requirement. Devices like digital cameras or cellphones that can make sudden power demands are better served with lithium-ion batteries.

9.5.3 Electric Vehicles

Now that we have a basic understanding of batteries, let's use them to build an electric vehicle. Our exemplar will be the 2014 BMW i3, shown in Fig. 9.17. Some of the characteristics of this car that we will need are given in Table 9.10.

The effective frontal area of the i3 is given by the product of its **drag coefficient** and the physical cross section. The lower the drag coefficient, the faster and more efficient the car. We must be careful to distinguish the available and rated energy of a lithium-ion battery because these can be permanently damaged if they are completely discharged. I have not been able to find values for the coefficient

---

[16]The energy is the charge times the rated voltage of the device.

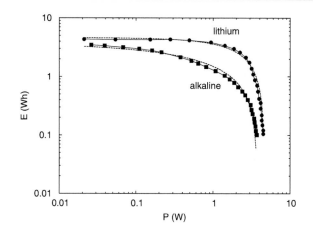

Figure 9.16: Power-energy curves for lithium and alkaline batteries.

Source: Q. Horn, Exponent Engineering and Scientific Consulting LLC.

Figure 9.17: 2014 BMW i3 electric vehicle.

of rolling friction (defined below) and the motor efficiency, so the numbers quoted are estimates. Notice that the specific energy of the i3 battery is 79 MJ/230 kg = 343 kJ/kg.

Our immediate goal is to determine whether electric vehicles can meet typical personal transportation needs. An acceleration of 0–100 km/h in 7.2 s is about the same as a Honda Civic, so this is within reason. The issue then becomes the maximum distance the car can travel. The rated distance for the i3, 160–190 km, certainly leaves room for improvement.

Table 9.10: BMW i3 characteristics.

Characteristic	Symbol	Value
Car mass	M	1200 kg
Car speed	v	–
Battery mass	M_b	230 kg
Effective frontal area	A_{eff}	0.7 m²
Range	R	160–190 km
Li-ion battery, rated	–	79 MJ
Li-ion battery, available	E_{batt}	68 MJ
Motor power rating	P	125 kW
Maximum speed	v_{max}	150 km/h
0–100 km/h	–	7.2 s
Coefficient of rolling resistance	C	0.01
Motor efficiency	ϵ	0.85

Let's build a model of electric vehicles to better understand these limitations. We focus on a car moving at highway speeds so that the energy consumption associated with starting and stopping need not be considered. In this case the motor must supply power to overcome the rolling resistance of the car and its air friction. Rolling resistance is due to internal friction in the car that generates a force that is proportional to the weight of the vehicle. The formula for this is CMg (look at Table 9.10 for definitions), where $g = 9.8$ m/s² is the acceleration due to gravity (at the Earth's surface). This is converted to a power by multiplying by the car's speed:

$$\text{power required to overcome rolling friction} = CMgv. \qquad (9.14)$$

The formula for air resistance is derived in exactly the same way as for wind turbines (look at Eq. 9.10). Thus

$$\text{power required to overcome air friction} = \frac{1}{2}\rho A_{eff}v^3. \qquad (9.15)$$

Recall that $\rho = 1.3$ kg/m³ is the density of air.

This power must be supplied by the car's motor, which has an efficiency of ϵ. Thus we derive

$$\epsilon P = CMgv + \frac{1}{2}\rho A_{eff}v^3. \qquad (9.16)$$

To determine the maximum range of the vehicle we equate the battery's total available energy to the power times the duration of the trip. Thus $E_{batt} = Pt$. But the time for the trip is R/v so we get

$$R = \frac{E_{batt}v}{P} = \frac{E_{batt}\epsilon}{CMg + \frac{1}{2}\rho A_{eff}v^2}. \tag{9.17}$$

Plugging in values from Table 9.10 gives the solid line in Fig. 9.18. Doubling the battery mass to 460 kg doubles the available energy (and increases the rolling friction since M goes up). The result is the dashed line. See Ex. 22 for the effect of this on the car's acceleration.

This calculation has assumed that the battery's capacity is independent of the draw rate, which is not true (Fig. 9.16). A fit to the lithium battery curve is shown as a dotted line in Fig. 9.16. I find that the energy available is approximately given by a linear relationship $E_{avail} = E_{batt} - cP$, where c is a coefficient with units of time that is specific to the battery. In the case of the figure $c = 1.0$ h. If we take this behavior into account the range will be diminished because greater power implies less available energy. The result for $c = 1000$ s is shown as dotted line in Fig. 9.18. The model predicts that the maximum range of the i3 is 160–170 km if one drives at 50 mph, which appears about right. The top speed of 150 km/h is shown on the x-axis in the figure. Notice that this nearly corresponds to the speed at which the battery is no longer capable of supplying energy.

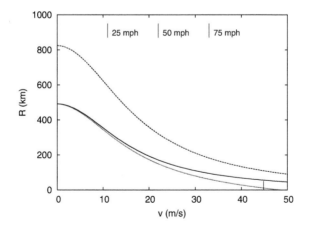

Figure 9.18: BMW i3 theoretical range.

9.5.4   A Practical Electric Transportation Scheme

It appears that getting a maximum range that corresponds to what people are ac-
customed to will be very difficult unless some breakthrough with battery design
is realized. It is possible that batteries with higher energy density will be invented
(lithium-ion batteries were unknown when I was young), but we should not rely on
this happening. Another option would be to design batteries that can be recharged
very quickly. Maybe stopping every 160 km would not be so bad if it only took
one minute to recharge your car.

I think another, radical but simple, solution exists. The idea is to use battery
power for local trips. Longer trips are made on highways, and these highways
should supply power to the vehicles on them. Say you want to go to Washington;
you drive to the local on-ramp of the highway where control of your car is passed
to an external computer. This computer runs all cars on the highway. Your car is
then driven onto the highway where it joins with other cars to make an efficient all-
electric virtual train with personalized cabins (i.e., your car). Power can be passed
to the trains via impedance or physical connectors. As you enjoy your coffee
while watching the scenery, your car batteries are recharged and your credit card
is debited appropriately. At the end of the trip the computer parks the car at an
off-ramp and you can take control for the last leg.

Ideally, cars will be self-driving in the near future so that one need not do any
driving at all. However, this relies on expensive detection and range-finding equipment. Highway travel is much simpler, so the artificial intelligence and sensing
required to make it feasible is minimal. Furthermore, human error will be removed from long distance travel, and hence such travel will become much safer.
Joining cars into virtual trains will save energy because the effective air resistance
per vehicle will be drastically reduced. And long distance trips will not be nearly
so tiresome.

Let's estimate the power required to run the national intelligent highway system. We will assume that one half of all power devoted to transportation goes
into long range travel. This is about 0.1 TW (for the electrified system). The National Highway System consists of 260,000 km of roadway; it carries 40 % of all
highway traffic and 75 % of all truck traffic. And 90 % of Americans live within
5 miles of the network.

The plan is to generate electricity on-site with photovoltaic panels. We assume
a typical insolation of 150 W/m^2 and 15 % efficiency so that 22 W/m^2 is available.
If a 20 m wide strip of PV panels is built alongside the highway (say on posts set
in the median), this will cover an area of 5.2 billion square meters which will be
capable of generating 0.11 TW of power. Of course, extra capacity will need to be
built or piped in to account for darkness and unforeseen demand.

Photovoltaic panels currently cost about $2/W; let us take the cost to be $10/W
with installation and control systems. The total cost to build the system would

then be about $50B per year spread over 20 years. This sounds stupendous, but is comparable to the federal highway administration budget and is less than 10 % of the military budget.

In terms of road construction, the cost to install the system would be around $7.2 million per mile. By comparison, the cost to build a four lane highway is $4–10 million per mile, so our figure is not ridiculous. However, the intelligent highway system will only require two lanes since computer control will permit dense traffic. This will realize a saving of $2–5 million dollars per mile, which goes a long way to offset the expense. It would be even better to replace asphalt with rail lines; this costs about $2 million per mile and permits even more efficient transport. And we should not forget the tens of thousands of lives that the new system will save every year.

Modern technology allows us to build a carbon-free, pollution-free long range transport system at an expense comparable to the current one.

One last detail needs to be addressed. Currently about 60 million cars are built per year in the world. If the future is more egalitarian one might expect one car for every four people; with a lifetime of 10 years, this would require building 175 million cars per year. A Nissan Leaf battery contains 4 kg of lithium. This battery is a little larger than the i3's so it is probably safe to use this as a typical requirement. Thus around 700 million kg/y = 700 ktLi/y are required in the electrified future. This is 20 times greater than the current production rate (Table 9.3), which means that the future production to reserves ratio is dropped to 18 years and the amount to production ratio is dropped to 100 years. In view of this, a robust manufacturer recycling program for old batteries seems prudent.

REVIEW

Important terminology:

anthropocene   [pg. 207]

DT and DD reactions   [pg. 226]

fusion reactor   [pg. 226]

linear and exponential growth   [pg. 202]

logarithm   [pg. 203]

logistics function   [pg. 207]

particulate matter, PM10, PM2.5   [pg. 219]

tokamak   [pg. 227]

Important concepts:

Growth is limited in population, the economy, resource exploitation, and exploration.

Air pollution has a negative effect on life expectancy.

Energy sources are solar, wind, hydro, carbon, fission, and fusion.

Electrification of the transport system is crucial to a carbon-free future.

FURTHER READING

G. Boyle, B. Everett, and J. Ramage, *Energy Systems and Sustainability*, Oxford University Press, 2003.

J. Diamond, *Guns, Germs, and Steel: The Fates of Human Societies*, W.W. Norton & Company, 1999.

D. MacKay, *Sustainable Energy – Without the Hot Air*, UIT Cambridge Ltd, 2009.

R. Wright, *A Short History of Progress*, Da Capo Press, 2005.

A large variety of governmental and UN information on energy and other resources is publically available. A short list is

Intergovernmental Panel on Climate Change, www.ipcc.ch

World Energy Council, www.worldenergy.org

US Geological Survey, www.usgs.gov

International Energy Agency, www.iea.org

EXERCISES

1. Fertility.

   Discuss the morality of having large families.

2. World Population Growth.

   What do you think caused the change in growth rate around the years

   (a) 5000 BC

   (b) 100 BC

   (c) 1950

   that can be seen in Fig. 9.3?

3. A Sun of People.

   Assume unlimited 2.2 % annual population growth, in what year would the mass of people equal the mass of the sun? Take the mass of the average person to be 75 kg.

4. Democracy and Population.

   Do you agree with the sentiment of Isaac Asimov quoted in Sect. 9.1?

5. Birth Rates.

   It has been claimed that birth rates fall as the education level of women rises. Does this sound plausible to you?

6. A Small World.

   Discuss ways in which the Earth is small and in which it is large.

7. The Anthropocene.

   Do you agree that we are in the "age of man"? In what ways can this idea be quantified?

8. Relative Production.

   Figures 9.7, 9.8, and 9.9 all show spikes when P is small. Can you find an explanation for this? BONUS: improve the logistics model to fit this behavior.

9. Discovery Rates

   Argue that the logistics model is applicable to resource discovery rates.

10. Drought.

    California is currently enduring a long-term drought. What options are available to Californians for dealing with this? What long term plans do you think should be made?

11. Reserve to Production.

    Table 9.3 presented many R/P ratios which are meant to estimate how long the resource will last. What is wrong with using R/P in this way? How can the estimates be improved?

12. Global Oil.

    Global oil production is about 100 Mb/d.

    (a) What do you think this means for global oil consumption?

    (b) A few times a year a discovery of "billions of barrels" of oil is trumpeted. What impact do these discoveries have on the global oil supply?

13. The Aral Sea.

    Look up the recent history of the Aral Sea. A living sea has been traded for cotton. Discuss the gains and costs in this transaction.

14. Natural Gas.

    The main component of natural gas in methane, CH_4. Use this to compute the mass of a cubic meter of natural gas.

15. Carbon Dioxide Emissions.

    (a) Supply the missing units in the computation in the Example of Sect. 9.3.5.

    (b) Obtain the result stated at the beginning of Sect. 9.3.5, 1 kg coal $\rightarrow$ 1 kg CO_2 in the atmosphere.

16. Carbon Dioxide and Temperature Change.

    Look at Fig. 9.11. Notice that the temperature change is essentially a linear function of added CO_2 for RCP8.5. The other scenarios are initially straight and then dip below the RCP8.5 curve. Explain why this happens.

17. PM2.5 I.

    In Sect. 9.3.6 it was mentioned that desert areas have high PM2.5 concentrations. Why is this?

18. PM2.5 II.

    (a) List some confounding factors that can affect the correlation seen in Fig. 9.14.

    (b) List confounding variables that can account for the increase in life expectancy the authors found in the 2000 data.

19. Hydro.

Does the sun or gravity provide the energy for hydroelectricity?

20. Home PV.

Research how much it would cost to buy and install sufficient PV panels to power your home or apartment. How does your result compare to your local power company's prices?

21. PV Costs.

Assume you are building a gigantic PV plant in the Sahara desert.

    (a) How large must the plant be to power the world?

    (b) Assuming a cost of $4.00/W, how much will it cost to build it? What expenses have been ignored in this estimate?

    (c) How must PV capacity must be added every year if the growth in energy demand remains 0.5 % and the world's population grows at 1 %/yr?

22. BMW i3 Acceleration.

Estimate the i3's time to accelerate to 100 km/hr if the battery mass has been doubled. Assume that motor characteristics remain the same.

23. BMW i3 Motor.

The BMW i3 has a 125 kW DC motor. What horsepower does this correspond to? Take efficiencies into account.

24. Future Generations.

Consider the following statement written by chemist Svante Arrhenius in 1925,

"Humanity stands . . . before a great problem of finding new raw materials and new sources of energy that shall never become exhausted. In the meantime we must not waste what we have, but must leave as much as possible for coming generations."

Do you agree with his sentiment? What arguments can be made for the other side?

25. Intelligent Electric Transportation.

Discuss problems with the electric transportation scheme of Sect. 9.5.4. What can be done to overcome these problems? Is it necessary to have PV farms along the highways? What benefits does this have?

Outlook

"The fault, dear Brutus, is not in our stars,
But in ourselves, that we are underlings."

— Shakespeare

We will close the book with a look at things on large scales. The first part of this chapter is a spatial "outlook" from Earth to the stars. It has long been tempting to imagine mankind leaving the smallish and grubby confines of Earth to float amongst the stars, settle new planets, and maybe even interact with alien species. Is this a fantasy or an eventual certainty? What about the more modest proposal of putting a colony on Mars? It's easy to imagine this colony, but is it practical or a death sentence?

The second section is a temporal "outlook" to the future. If we are indeed trapped on Earth, what is in store? How much do our actions matter? And what will future civilization look like?

10.1 Space Travel

You might be surprised to see a discussion of space travel in this book. I have included it because some enthusiasts imagine an unfettered Star Trekish future for mankind (Fig. 10.1); hence worries about resource management, pollution, climate change, and population growth are all misplaced. The root issue is that it is easy to *imagine* space travel – you just get into your rocket and wait for a while and then you arrive at your new planet. While we are at it, the planet may as well be Earth-like and inhabited by attractive aliens.

10.1.1 To Mars!

Let's step back and look at a much simpler problem: colonizing Mars. The euphoria surrounding the recent Rover missions to Mars apparently inspired many people to dewy-eyed dreaming. One op-ed in the New York Times said,

© Springer International Publishing Switzerland 2016
E.S. Swanson, *Science and Society*,
DOI 10.1007/978-3-319-21987-5_10

Figure 10.1: Space colony fantasies.

> "But Mars is waiting. It spins now outside our human reach. We must
> realize that the work of growing up is not something we can cut when
> the budget gets tight. It is mission critical, for the intellectual life of
> the species, for the future of humans, not to stagnate, not to wither,
> but to stretch, and reach, and always to expand."

The enthusiasm of the writer is palpable. Surely we are doomed if we are
trapped on our paltry planet! Our intrepid op-ed writer is not the first to be en-
tranced with such ideas:

> "When I trace, at my pleasure, the windings to and fro of the heavenly
> bodies, I no longer touch Earth with my feet. I stand in the presence
> of Zeus himself and take my fill of ambrosia."
>
> — Ptolemy, *Almagest*, c. 170.

But no amount of enthusiasm can overcome the laws of physics. Nature, quite
simply, does not care what we think.

Our first job is to get to Mars. As they slip around their orbits, the distance between Earth and Mars varies between 56 and 400 million km, so it is going to take some time to get there. Recent probes have averaged about 200 days to make the trip.

Breaking free from Earth's gravitational field is not easy. In fact a spacecraft of the size necessary to get to Mars cannot be built on Earth and must be ferried piecemeal into low Earth orbit and assembled there. NASA estimates that the craft would have a mass of 1250 t. By comparison the International Space Station has a mass of 419 t and took 10 years and 115 flights to build. Unless things change dramatically (like new heavy lift rockets are developed), we are looking at 30 years to construct a spacecraft capable of going to Mars. Current costs to carry 1 kg into low Earth orbit is around $10,000, so just ferrying the bits of the Mars rocket will cost more than $12 billion. The total cost of the ISS is estimated to be $150 billion, and it is arguably the most expensive thing ever built. Based on this, a naive estimate is that a Mars rocket will cost $½ trillion. This is just the cost of one ship; it is likely several will be required for any colonization effort.

Once assembly is complete and the crew is settled in, the journey can begin! But we need to be wary; the Earth's atmosphere and magnetic field protect us from high energy solar photons and cosmic rays. It is estimated that a 200 day trip to Mars will expose each astronaut to 330 mSv of radiation. As a reminder (check Table 8.2), this is 100 times higher than the average natural dose and close to the threshold of radiation poisoning. Clearly the rocket is going to have to be shielded to keep the astronauts from frying.

The troubles are not over; it turns out even landing on Mars is difficult. The problem is that the Martian atmosphere is too thin for parachutes to work. Retrorockets could be used but the atmosphere is thick enough that turbulence can destabilize the descent. Even igniting a rocket on a spacecraft that will be traveling much faster than the speed of sound is difficult. Quite simply, the technology to land craft heavier than 1 tonne has not been developed yet.

Six or seven months in zero-gravity has a debilitating effect on people. The main problems are loss of muscle and bone mass. In an effort to combat muscle loss, the ISS crew does an hour of cardiovascular training and an hour of weightlifting every day. Even then, they struggle to walk after a 6 month mission. Bones also wither when removed from the stresses caused by gravity. Load-bearing bones can lose as much as 1–2 % of their mass per month. The calcium contained in the former bone ends up in the bloodstream where it wreaks havoc by causing kidney stones, constipation, and even psychotic depression.

10.1.2   Living on Mars

Getting to Mars may not be easy, but things will be better once we get there, right? Mars, it turns out, is not a very pleasant place. Its atmosphere is thin and there is no protecting magnetic field, so radiation doses will continue to be high. One estimate is that colonists will receive 200 mSv doses every year – 70 times higher than on Earth. The atmosphere contains no oxygen and is so rich in carbon dioxide that it is toxic to animals and plants. And the temperature varies between −87 and −5 C. In short, no one will be going outside for long strolls. People will need to live permanently in shielded airtight containers, always afraid of the slightest breach of compartment integrity.

The colonists will need food, water, and air once they are safe in their containers. We now build a simple model of a human, illustrated in Fig. 10.2, so that these requirements can be quantified. The model assumes that our basic needs can be simplified to the substances glucose, nitrogen, water, and oxygen. The first two will comprise "food". Oxygen is required to oxidize "food" and produce the energy that runs the body and allows us to do work (otherwise known as "waste heat"). Specifically, glucose oxidization is the process

$$C_6H_{12}O_6 + 6\,O_2 \rightarrow 6\,CO_2 + 6\,H_2O + \text{ energy.} \qquad (10.1)$$

The excess carbon dioxide produced in this process is expelled by our lungs and the released energy is 2870 kJ for every mole of glucose plus 6 moles of O_2. Six moles of O_2 is equivalent to 134 L, thus 21 kJ (= 5 Cal) of energy is released per liter of oxygen gas. Alternatively, one liter of oxygen gas is required to metabolize one gram of sugar or 0.5 gram of fat (refer to page 92 for help with getting these numbers). These figures tell us that producing 2000 Cal of energy per day requires 400 L = 0.4 m^3 of O_2 per day.

Nitrogen is also absorbed by the body and is metabolized to produce energy and ammonia. Unfortunately ammonia is toxic, so the body invests some energy to convert it into benign **urea**. This is filtered out (along with salts and other trace nasties) by the kidneys and flushed from the body as urine.

Finally, water is needed for flushing urea (about 1.5 L/day) and because we are rather leaky – losing 0.5–0.8 L of water per day via exhaled humid air (life on Mars will be too sedate for sweat or tears).

Our model indicates that water does not need to be produced for the colonists because it can be recycled from air and urine. Oxygen will be consumed at the rate of 0.4 m^3/person/day so a steady source will be required. Fortunately, oxygen can be made easily by **electrolysis**, which is the process

$$\text{energy} + 2\,H_2O \rightarrow 2\,H_2 + O_2. \qquad (10.2)$$

The computations above tell us that people require $400/134 \cdot 6 = 18$ moles of O_2 per day. Electrolysis requires 286 kJ per mole of water, thus using this process

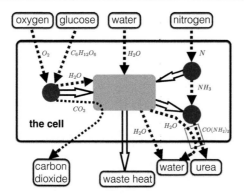

Figure 10.2: A simplified human. *Large arrows* represent energy flow. *Dotted lines* are material flow. *Shaded circles* denote chemical processes.

to generate sufficient oxygen for 1 day of breathing will require 18 mole · 2 · 286 kJ/mole/day = 120 W of power per person (the extra factor of 2 is because 1 mole of water only makes 0.5 moles of O_2). This modest power demand is good news. The bad news is that, unlike water, oxygen does not form a cycle. What is missing, of course, is plants, which can convert carbon dioxide into oxygen via photosynthesis.

Let us imagine a less grandiose attempt at establishing a Mars colony. These actually exist now – a company called Mars One proposes to send crews of four people on one-way trips to Mars every 2 years starting from the year 2024. They claim that 200,000 people have applied to be colonists so far (Fig. 10.3).

In the Mars One plan, water for external use and electrolysis will be collected from water-bearing soil. Power will be provided by 3000 square meters of photovoltaic panels, which should supply 300–600 kW.[1] This must supply power for electrolysis, lighting, soil collection, waste removal, and heating. It might all be feasible, but Mars is a harsh environment and any breakdowns would certainly mean death for everyone concerned.

Food production is the weak link in the Mars One plan (in fact no plan for growing food is mentioned on their web site). Martian farming is not straightforward; plants, after all, evolved on Earth just as much as humans. Colonial plants will require a radiation-safe environment with hydroponics, CO_2, water, fertilizer, and lighting. There are two options: you can grow food in a fully pressurized environment, or you can grow it in a low-pressure environment.

The second option has the great benefit of being more economical because it is difficult to keep atmosphere in when it is placed under pressure. The problem

---

[1] It would be possible to transport miniature nuclear reactors to Mars. These are currently used to power submarines and development of small commercial units is underway. The smallest reactor that I am aware of powered the experimental naval submarine NR-1 ("Nerwin"). I estimate it weighed 50 t and produced 25 MW of power.

Figure 10.3: A Martian death trap? The Mars one colony.

Reproduced with permission, Bryan Versteeg/Mars One.

with this approach is that plants have evolved under Earth pressures and it is not certain how they will respond to a low pressure conditions. Recent experiments indicate that plants under low pressure think they are under drought stress because water is pulled through them easily. The resulting drought defense mechanisms severely impact plant growth.[2]

Because of issues such as these, it is preferable to grow plants at normal (Earth) atmospheric pressure. But this means building large pressurized greenhouses. How large these must be can be estimated by colonial energy needs. From Table 4.3 we see that the most energetic foods supply about 100 GJ/ha. Assuming 2000 Cal/day/person are required and two crops per year are available, we find that $150\,m^2$ of high-yield crop is required per person. This is a sizable enclosure that must be transported to Mars at enormous expense.

The crops also carry a substantial power demand. Since they supply 97 W per person they must require at least this in power from light. In fact photosynthesis is only 0.1–10 % efficient, depending on the crop. Assuming a 1 % efficiency means that 10 kW power must be supplied in the form of light (assuming complete coverage from the Martian sun).

Keeping the plants from freezing is more of a concern. Convection losses are not a problem because they are in a greenhouse. Also heat loss by direct conduction is not significant because the Martian atmosphere is very thin. This leaves thermal radiation as the primary mechanism of heat loss. The power needed

---

[2]Source: A.-L. Paul et al., *Hypobaric Biology: Arabidopsis Gene Expression at Low Atmospheric Pressure*, Plant Physio. **134**, 215 (2004).

to heat the greenhouse to overcome this loss can be estimated with the Stefan-Boltzmann law (Sect. 6.3). If we call A the surface area of the greenhouse then

$$P_{\text{heat}} = \sigma A T_{\text{inside}}^4 - \sigma A T_{\text{outside}}^4 = 5.67 \cdot 10^{-8} \cdot 150 \cdot (293^4 - 193^4) = 50\ \text{kW}.$$

Of course, inefficiencies in the heating system should also be taken into account. All told, power requirements will probably top 100 kW/person, so the planned Mars One PV array will only support 3–6 colonists.

Building a large-scale Martian city is even more difficult to attain. Large structures require a massive building program, which requires a massive mining, smelting, and processing program, which requires vast amounts of energy. The only viable way to build up this infrastructure is with enormous numbers of autonomous robots which would have to be sent regularly to Mars over a period of many of years.

These robots would need to be able to smelt minerals, manufacture computer chips, build fusion or fission reactors, and repair themselves. In short, they must build an entire functioning and sophisticated robotic society simply to prepare the way for the weak meatbags.

At this stage you might wonder why we would bother going to Mars. There would be nothing to do but watch robots work for us. And the robots might not take kindly to that.

10.1.3 To the Stars!

Forget about Mars! Let's go someplace good. Just like Earth, but with more beaches. Wherever this place is, it is far away. Proxima Centauri is the nearest star, at 4.24 ly ("ly" is a light year, which is the distance light can go in 1 year). It is 10 ly to the nearest known exoplanet and most of them are more than 100 ly away. The closest Earth-like exoplanets that I know about are Kepler-62e and Kepler-62f, which are 1200 ly away.[3]

Let's be generous and assume that we only need to go 100 ly to find our new home. That is still a long way, so let's assume that the spacecraft moves at some fraction of the speed of light, $v = fc$. The speed of light is the fastest anything can move so $0 < f < 1$. We also don't want f to be too small since the trip will take too long. How big must the ship be? If the colonists grow their own food, its going to need about 100 m³/person. The density of the ISS is about ½ t/m³, so we'll need 50 t/person. Let's call it 100 t/person to be safe.

With this mass, the energy required to move at speed $v = fc$ is

$$\frac{1}{2}mv^2 = \frac{1}{2} \cdot 10^5 \cdot f^2 c^2 \approx f^2 10^{22}\ \text{J/person}. \qquad (10.3)$$

---

[3]Source: W.J. Borucki *et al.*, *Kepler-62: A Five-Planet System with Planets of 1.4 and 1.6 Earth Radii in the Habitable Zone*, Science **340**, 587 (2013).

Unless f is tiny, this represents a vast energy. For example, a fuel that supplies 100 MJ/kg (gas is 42 MJ/kg) means that $f^2 10^{14}$ kg/person of fuel is required. This dwarfs the mass of the space ship that we assumed, so that calculation needs to be redone with a bigger mass, which means higher energy, which means yet more fuel. Clearly this is an impractical way to get to our new home.

A better approach would be to generate energy with a nuclear reactor (recall that fusion reactors must be huge, so let's stick with fission). The *minimum* power required for the trip is $1000 f^3$ GW/person. A full scale commercial nuclear reactor generates about 1 GW of power, so a fast trip would require 1000 nuclear reactors per person. A slow trip with $f = 1/100$ would only require one reactor per 1000 people, but would take 10,000 years to complete. Of course these reactors add their own mass to the ship and require fuel.

In Sect. 9.4.4 we learned that 400 1 GW reactors require 66 kt of uranium per year. Thus the ship's $1000 f^2$ reactors/person will require mining $10^{10} f^2$ kgU/p. A 1 GW reactor uses about 15 t of low enriched uranium per year, so $1.5 \cdot 10^9 f^2$ kg LEU/person would be required to power the voyage. Again, our ship is being swamped by its energy requirements.

It is possible to do a little better with energy estimates by deriving two simple equations. We will model the rocket with three components: a living area with mass L, a power generation area with mass R, and a propellant with mass P. A propellant is needed because a ship can only move forward by throwing something backward. In a conventional rocket this is the exhaust from burning chemical fuel; in our case it can be anything convenient that can be accelerated with the ship's technology.

Imagine a ship at rest ejects some propellant of mass δP (δ means "small") at speed u. Conservation of momentum (momentum before = momentum after) and the equation, `momentum = mass · speed`, tell us that the ship has to recoil (i.e., move forward) with a speed given by the ***rocket equation***:

$$M \delta v = u \delta P. \tag{10.4}$$

Uniformly ejecting propellant, as imagined here, implies a gradually increasing acceleration of the ship since it gets lighter as time goes by. This is inconvenient since one would prefer that the colonists experience an acceleration equal to gravity. We therefore imagine that ship engineers arrange this to occur for one half of the voyage. At the half way point of the trip the rocket can be rotated 180° so that it will decelerate, eventually coming to a halt as the ship approaches its destination.

The next task is to figure out how much power it takes to eject the propellant. First, let's call the specific power of the ship's reactor, ρ (recall that this is measured in W/kg). The required power can be obtained by conserving energy.

Namely, the energy supplied by the reactor in the time δt must equal the kinetic energy of the propellant:

$$\epsilon \rho R \delta t = \frac{1}{2} \delta P u^2. \tag{10.5}$$

Here ϵ represents the efficiency for converting reactor energy into propellant energy. The rate at which propellant is being ejected is $\delta P / \delta t$, which we call $\dot{P}$. The last equation implies $\dot{P} = 2\epsilon \rho R / u^2$. According to Eq. 10.4 the acceleration is given by

$$a = \frac{\delta v}{\delta t} = u \frac{\dot{P}}{M} = \frac{2\epsilon \rho R}{Mu}, \tag{10.6}$$

where we have substituted for $\dot{P}$ as given above and for $\delta P / M$ as given in Eq. 10.4.

Remember that M is the total mass of the ship, given by $R + L + P$. This is less than $R + L + P_0 = R + L + 2\epsilon R \rho T / u^2$, which is less than $R + 2\epsilon R \rho T / u^2$. The symbol T, representing the total time of the trip has been introduced. Thus

$$a < \frac{2\epsilon \rho R}{uR + 2\epsilon R \rho T / u}. \tag{10.7}$$

The maximum acceleration that can be achieved occurs when $u = \sqrt{2\epsilon \rho T}$, which implies

$$a_{max} < \sqrt{\frac{\epsilon \rho}{2T}}. \tag{10.8}$$

Say we want to accelerate (or decelerate) the ship at $a = 9.8\,\mathrm{m/s^2}$ for a total period of 100 years.[4] Then the specific power required by the ship's propulsion system is

$$\rho > 6 \cdot 10^{11}\ \mathrm{W/kg}. \tag{10.9}$$

How does this figure compare to current technology? Nerwin's 50t nuclear reactor produces 25 MW of power, yielding $\rho = 500\,\mathrm{W/kg}$. We are *nowhere near* the required specific power needed. Again, simple energetics appear to be telling us that interstellar travel is extremely difficult, if not impossible.

Space enthusiasts will know that I have been neglecting an important wrinkle in our analysis of interstellar travel. This is Einstein's **Theory of Relativity**, which modifies familiar laws of physics when motion is fast. It would take us too far afield to derive all of Einstein's equations so we focus on a few main concepts. The main result of the theory is that

---

[4]This means the ship must be moving very close to the speed of light. How plausible this is will be discussed shortly.

distance and time are relative to the observer.

This means that space and time can no longer be thought of as separate things, but must be considered as aspects of one thing, called **spacetime**.

One of the famous implications of Einstein's theory is that time as measured by a moving observer runs slower than measured by a stationary observer. In other words, if you are moving very quickly (with respect to me) and an hour ticks by on your clock, more than an hour will have gone by on my clock. This has obvious important implications for interstellar space travel with fast ships.

The equations for relativistic motion under constant acceleration a (as measured on the spacecraft) are given in terms of t (the time as measured on Earth), and v (the spacecraft speed as measured by observers on Earth). In this case

$$at = \frac{v}{\sqrt{1 - v^2/c^2}}, \tag{10.10}$$

which can be solved to obtain

$$\frac{v}{c} = \frac{at}{\sqrt{c^2 + a^2 t^2}}. \tag{10.11}$$

Running clocks do indeed run slow: if the colonists measure an elapsed ship time of T then on Earth this will appear as

$$T_{\text{Earth}} \approx \frac{c}{2a} e^{aT/C}. \tag{10.12}$$

This equation is accurate for large times or accelerations. Lastly, the time necessary to travel a distance x is given by

$$t = \frac{1}{ac}\sqrt{a^2 x^2 + 2ac^2 x} \approx \frac{x}{c}. \tag{10.13}$$

We are now ready to check some of the previous assumptions. A steady acceleration that is maintained for a long time will bring the speed of the ship very close to the speed of light. Equation 10.11 tells us that $f = 0.99975$ when an acceleration of 9.8 m/s^2 has been maintained for 50 years. Equation 10.13 then tells us that the first leg of the trip will last about 50 years. Thus the assumptions made above are reasonable. Lastly, Eq. 10.12 indicates that the elapsed time for the entire trip as experienced on the spaceship will be around 9 years.

What about the energetics? Although Einstein's theory of relativity makes it feasible to do the journey in a reasonable time (if you regard 100 years as reasonable), it makes the energetics worse. The problem is that the speed of light is the maximum speed attainable in Einstein's theory, and it takes an ever increasing amount of energy to push the spacecraft's speed closer to c. The end result is that an even greater specific power is required.

Humanity is tethered to the Mother Planet.

10.1.4 A Practical Scheme for Space Exploration

It appears that a fast trip to the exoplanets is out of reach. How feasible is a slow trip? In this case we need not worry about Einstein's equations and relativity and can stick with Newton's equations.

Not seeking to be modest, let us power the slow trip with a commercial-scale nuclear reactor. This supplies 1 GW power, and with a power density of 500 W/kg, implies that 2 kt of material needs to be transported to low Earth orbit for assembly. The shuttle could lift 24 t to low Earth orbit, so this will take about 100 shuttle flights over 10 years at a cost of around $20 billion (just for the transportation). Since this seems about the limit of what is practical, let's assume that the "living" area and the propellant also amount to 2 kt each. The ISS weighed in at 419 t and cost $150 billion to build, so a naive estimate for the colonial spacecraft cost is $2 trillion. If this sounds like a lot, consider that the world spends $1.5 trillion per year on its military.

The plan is to use half the propellant to accelerate to a final speed v_f, coast at this speed for a time T_c, and then use the last of the propellant to come to a halt at the destination. If the time to reach v_f is called T_1, then the total time for the trip is $2T_1 + T_c$. The equations we need are

$$\epsilon \rho R = \frac{1}{2} \frac{P_0}{2T_1} u^2 \tag{10.14}$$

and

$$v(t) = u \log \left(\frac{R + L + P_0}{R + L + P_0 - \frac{P_0}{2T_1} t} \right). \tag{10.15}$$

The first of these is analogous to Eq. 10.5 and the second can be derived from the rocket equation.

The distance traveled while accelerating is given by integrating the last equation and is (using the assumptions for L, R, and P_0)

$$D_1 = u T_1 \cdot (1 + 5 \log(5/6)) = 0.088 \, u T_1. \tag{10.16}$$

The terminal speed can be obtained from Eq. 10.15 by substituting $t = T_1$ and is $v_f = u \log(6/5)$. Thus the total time for a trip of distance D is

$$T = 2T_1 + \frac{D - 2D_1}{v_f} = \frac{P_0 u^2}{2\epsilon \rho R} + \frac{D}{u \log(6/5)} - 2\frac{0.088\, u T_1}{u \log(6/5)} \tag{10.17}$$

or

$$T = \frac{0.51\, P_0 u^2}{2\epsilon \rho R} + \frac{D}{u \log(6/5)}. \tag{10.18}$$

This is an expression for the total time in terms of the propellant ejection speed, u, which can be minimized by choosing

$$u = \left(\frac{D\epsilon \rho R}{0.51 \log(6/5) P_0} \right)^{1/3}, \tag{10.19}$$

which gives

$$T_{\min} = \left(\frac{0.51 D^2 P_0}{2 \log^2(6/5)\epsilon \rho R} \right)^{1/3} \cdot \left(2^{-2/3} + 2^{1/3} \right). \tag{10.20}$$

Plugging in numbers[5] gives $u = 0.044c$, $v_f = 0.008c$, $T_1 = 5600$ years, and

$$T_{\min} = 18{,}500 \text{ years.} \tag{10.21}$$

The slow trip really will be slow, but it is feasible and might actually work!

Maintaining life on a ship, no matter how big, for 18,500 years is not simple. It would be more sensible to eliminate all that life support infrastructure. Alternatively, running a reactor for 18,500 years (or anything for that matter) is a daunting task. It is all too easy to imagine a nice shiny space ship leaving Earth, and an equally shiny spaceship arriving at its destination 18,500 years later. But is this realistic? My car, after all, is a scant 13 years old and is in the shop at least three times a year. Thus it is apparent that the ship will need to be "manned" by some very sophisticated, intelligent robots. These robots will need to be able to diagnose and repair problems as they arise. If we insist that people go along for the ride, maybe they can be carried as frozen embryos, ready to be thawed and raised by robotic mothers after their 18,500 year slumber.

If we are going to explore beyond the solar system it must be by proxy.

Again, it is important to be practical. The odds of any piece of machinery lasting so long are essentially zero. The ship would thus need to be equipped with

---

[5]$D = 100$ ly, $\epsilon = 0.5$, $R = L = P_0 = 2\,$kt, $\rho R = 1\,$GW.

the infrastructure necessary to create computer chips, repair nuclear reactors, and perform a myriad of other high technology chores. None of this sounds simple, and it would not be surprising if it were essentially impossible to achieve. In any case, it is evident that the robots will either incorporate human values in their programming – or perhaps could have actual consciousness loaded into them. In view of this, will it really be important that the wet biological aspects of humanity travel along?

10.2 Looking Out

10.2.1 The Anthropocene

We are living in a unique period of irrational exuberance. The discovery of the New World 500 years ago opened up a vast resource-filled territory ripe for exploitation. At the time, the main source of energy was human labor, and this had to be imported to permit resource exploitation – a fact that has societal repercussions to this day. The exuberance only accelerated when the industrial revolution – coupled with ready and cheap energy sources – permitted even more rapid exploitation of resources. The net result is an illusion that rapid growth is a natural state of man. But it is not. The Paleolithic era (old stone age) lasted from 3 million years ago to 12,000 years ago. From the first chipped stone tool to the development of iron tools took 2 million years. But it only took 3000 years to go from that first primitive iron implement to the hydrogen bomb.

We have studied some of the consequences of this exuberance. Pumping carbon dioxide into the atmosphere feeds our energy addiction at the price of warming the planet, raising sea levels, acidifying the oceans, and reducing our life spans. Water is being withdrawn at unsustainable rates. The economy is run with reckless disregard for the waste we produce – as if the planet were simultaneously a limitless resource and cesspool. As a by-product, the population has exploded from 2 to 7 billion people in the past 90 years. We have caused the extinction of so many species that our dominion over the earth will appear like the impact of an asteroid in the fossil record.

History supplies us with countless examples of how this mind-set can lead to disaster. Easter Island is surely the classic exemplar of man's penchant to wear out his welcome with nature. The island, 3500 km off the coast of Chile, was settled by Polynesians about 1000 years ago. Although they created a thriving culture (emplacing hundreds of the famed moai statues), overpopulation led to deforestation, and hence the ability to build boats. By the time of European contact in the 1700s the population had dropped by 80 %. Ensuing devastation by tuberculosis, smallpox, and kidnapping by slavers reduced the population to 111 individuals (Fig. 10.4).

It takes a lot to kill a city. Even the devastation wrought by Little Boy was not enough to wipe out Hiroshima. The only known way to kill a city is to wreck its economy. Half-buried ruins around the world attest to our ability to do just this.

Figure 10.4: Moai on Easter Island.

Photo credit: Jim Richardson.

Ironically, it is the economic success of the city that leads to population pressure, which all too often leads to unsustainable resource extraction, and collapse. Even entire civilizations can suffer this fate. A prime example is provided by the Sumerians, who flourished from 4000 BC to 2000 BC in southern Iraq. It is thought that salt poisoning of the land brought about by excessive irrigation was a major contributor to the Sumerian downfall.

> "We're Ice Age hunters with a shave and a suit. We are not good long-term thinkers. We would much rather gorge ourselves on dead mammoths by driving a herd over a cliff than figure out how to conserve the herd so it can feed us and our children forever. That is the transition our civilization has to make." – Ronald Wright

10.2.2   Collective Action

Can anything be done? To answer this we must face two important facts:

  (i) We are stuck on Earth for a very long time, and possibly forever. We therefore cannot continue with old habits and simply move on when we befoul our habitat.

 (ii) Market forces are incapable of dealing with the coming slow motion crisis.

Free enterprise is good at bringing us a new cell phone every year, but it is simply not designed for stewardship because there are no incentives for sustainable development. If you are making money from harvesting dodos or pumping oil there is no reason to do so in a sustainable manner. Rather, profits are maximized by exploiting the resource as quickly and effectively as possible. If you destroy the resource as a result, you simply take your windfall and invest it in something else.

What is required is societal change. We have demonstrated many times that society is capable of making rapid changes in response to new challenges. The notion of animal rights did not exist 200 years ago and it was perfectly acceptable, for example, to whip a horse to death. But a heightening awareness of pervasive cruelty led to laws protecting animals from abuse (starting in 1835 in England[6]), which have spread around the world with wide support.

One hundred years ago it was acceptable for company owners to hire mercenaries to carry out pitched battles with strikers.[7] And consumers could be sold questionable or even dangerous products with no restrictions. But catastrophes such as the Triangle Shirtwaist Factory Fire (1911), the Cocoanut Grove Fire (1942), and countless mine disasters eventually led to the establishment of fire code and worker and consumer protection acts. More recently, antismoking, civil rights, gay rights, and species protection campaigns have all had some degree of success and point to a healthier and more equitable future for all.

As important as all these social campaigns were, it is the environmental movement that is most relevant to this discussion. Environmentalism had its roots in the industrialization of the nineteenth century when a growing concern was raised over rampant air pollution. In England, the Coal Smoke Abatement Society was formed in 1898 because of the famous "London fog" caused by burning coal. In America, John Muir (1838–1914) and others advocated for the creation of national parks and preserving natural places. Muir helped establish the Sierra Club in 1892, which continues to advocate for a variety of environmental issues today.

Despite this early work, environmentalism did not catch on with the general public until Rachel Carson (1907–1964) published *Silent Spring* in 1962. Her book documented the serious consequences massive spraying of the pesticide DDT was having on the environment. Not surprisingly, Carson's book was denounced by the chemical industry. However, her claims were upheld by subsequent governmental investigations and DDT was banned in the United States in 1972. Similar narratives have surrounded a number of other chemicals, with wide untested use followed by suspicions of negative effects, scientific testing, denial and often unscrupulous smear campaigns, and finally restrictive legislation. Examples within memory include mercury, leaded paint and gas, thalidomide, asbestos, acid rain, and chlorofluorocarbons (Fig. 10.5).

---

[6]In a telling chronology, slavery was only abolished 2 years earlier in England. It would be another 12 years before it was abolished in Pennsylvania, 30 years in the United States, 53 years in Brazil, and 127 years in Saudi Arabia.

[7]I refer to the famed Homestead Strike of 1892.

Figure 10.5: Rachel Louise Carson

The story of the detection and amelioration of the ozone hole provides a recent and compelling example of this narrative. Ozone (O_3) is a rare component of the atmosphere that is most prevalent in the lower stratosphere (10–15 km up). This layer is of vital importance because it absorbs at least 97 % of dangerous solar UV radiation (UV-A largely passes through to cause tanning and skin cancer). Ozone can be destroyed by catalytic cycles[8] that contain chlorine.

In 1974 the chemists Frank Rowland (1927–2012) and Mario Molina (1943 -) claimed that chlorofluorocarbons (CFCs) can destroy ozone. This was a concern because CFCs stay in the atmosphere a long time (subsequent analysis showed that *virtually all* CFCs ever emitted were still in the atmosphere), and they were in common use as aerosol propellants and refrigerants. However, the problem was not regarded as urgent, in part because a National Academy of Sciences report estimated that global ozone would only be depleted by 5–9 % in 100 years of (growing) CFC use. Nevertheless, the governments of the United States, Canada, and Norway took action and banned CFC aerosol spray in 1978 (in a rare volte-face, the European Union refused to act).

Only 3 years later Joe Farman, Brian Gardiner, and Jonathan Shanklin announced that atmospheric ozone concentration had dropped by a startling 30 % in the Antarctic in late fall. Within a few years, ensuing intense research was able to explain why the drop in ozone was so rapid, why it happened in Antarctica, and why it happened in the fall.[9] In September of 1987, 24 countries signed the

---

[8] A catalyst is a substance that can increase the rate of a chemical reaction without undergoing any change itself.

[9] The explanation involved large scale atmospheric currents and catalysis precipitated by airborne ice crystals.

Montreal Protocol which set in motion an effort to eliminate CFCs entirely by the year 2000. Eight years later Frank Rowland and Mario Molina were awarded the Nobel Prize in chemistry.

There is little doubt that the vigilance of a few scientists and the quick actions of many governments averted a catastrophe. Figure 10.6 shows the predicted prevalence of ozone with current practices in the top row and what would have occurred if CFCs had not been banned in the bottom row. The "world avoided" of 2060 is rather frightening, with essentially no protective ozone over the Antarctic and three times less ozone than currently present over the rest of the world.

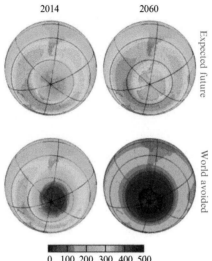

Figure 10.6: Ozone concentrations, real and avoided.

It is tempting to regard the CFC saga as a warning on human hubris and unintended consequences. But this is not the only message to take away. The ecosystem is complex and not all consequences can be accurately foreseen. Thus vigilance is of prime importance. At the other extreme, wilfully shutting one's eyes to evidence and the scientific process should be strenuously rejected. Thus actions like cancelling NASA's Deep Space Climate Observatory appear to be shortsighted.[10]

Equal vigilance over societal actions is required. Certainly the current agitation against cellphones and vaccines is based on scientific rubbish. More seri-

---

[10]I am happy to report that the satellite was launched in 2015 after twelve years in storage.

ously, the anti-nuclear power positions taken by the Sierra Club and the Union
of Concerned Scientists strike me as shortsighted and alarmist. The contrast between the public (and hence governmental) attitude toward fine particulate matter
and radioactivity could not be sharper. Radioactivity, which occurs at a natural
level of 3 mSv/yr, is heavily regulated while its attendant public health risks are
evaluated with the uncompromising linear no-threshold model. This is in spite
of demonstrated negligible negative health outcomes, even after the worst of disasters (Sect. 8.7). In contrast, industry continues to pour particulate matter into
the atmosphere *even though we know it shortens lives*.[11] Applying the linear no-
threshold model to this pollutant reveals a cost of millions of lost life-years. And
yet we tolerate it.

10.2.3   Thinking Ahead

> "Progress is impossible without change, and those who cannot change
> their minds cannot change anything."
> — George Bernard Shaw.

The success stories just described should provide succor for future (and current) generations. With sufficient motivation and urgency, it is possible for the
public to force politicians to act.

Compared to other problems, resource management should be straightforward. Slowly but surely, helium will be permanently lost. There is little we
can do about this except try to not be wasteful. Wasting a good portion of the
global supply on balloons is obviously foolish; banning this frivolity will be a
good bellwether for society's (growing?) maturity.

Political and economic pressure cause many regions to withdraw fossil water
at unsustainable rates. Because I see little hope for reining this behavior in, the
best course of action will be to plan for arid futures in the American southwest,
north Africa, the Middle East, northern India, and northern China. In practice this
will mean drawing down population in these areas. We can do it, or Nature will
do it for us.

It is likely that unregulated market forces will be able to handle the coming
crunch with other resources. At some point it will simply become too expensive
to find new gold or copper and these minerals will have to be efficiently and effectively recycled. In short, gold and copper and the rest will join water, carbon,
and nitrogen in making an *ecological cycle*. We will thus merely borrow minerals
for a while before they are returned to the great planetary cycle.

Our carbon addiction is the most pressing current problem. Our addiction
is warming and polluting the planet, with possibly serious repercussions. Sea
levels may rise between 1 and 7 meters, flooding valuable delta farmlands that
support hundreds of millions of people. The oceans are well on their way to a

---

[11]Recall that each additional $10 \, \mu g/m^2$ of PM2.5 shortens life spans by 7 months.

more acidic future. What effect will this have on marine life? At a personal level, the car I drive emits particulate matter that shortens the lifespan of my family. But the connection is not immediate, so we continue with our slow and steady self-destruction.

The simple solution is to do what we have done many times in the past with pollutants: stop. This was not so difficult for CFCs or mercury, but carbon makes us rich, and we like that. It would thus be prudent to develop alternative forms of energy. As we have seen, much is being done, but more is needed. Table 10.1 presents a summary of some of the conclusions from Chap. 9. Recall that the current world average power demand (*very* unequally spread) is 2300 W/person.

Table 10.1: Current and projected carbonless power supply.

source	Current (W/person)	Projected (W/person)
Hydro	125	350
Wind	45	400
Solar	20	380
Nuclear	53	150
Fusion	0	?
Total	243	1280 + ?

As can be seen, about $1/10$ of the current global energy demand is met by carbon-free sources. The third column of the table shows projections. It is estimated that unutilized hydropower resources, primarily in Africa, can contribute 350 W/person. Given that this resource is renewable and relatively environmentally benign, it seems morally imperative that it be exploited where possible.

Recall that wind power requires large areas; we estimated that 12 % of the world's surface area must be covered to reach 2300 W/person. This appears unreasonable. However, wind turbines are not excessively intrusive (except to our field of view) and it might be possible to cover 2 % of land area. This gives the figure of 400 W/person.

Photovoltaic panels provide about twenty times the power density of wind turbines, and therefore make fewer demands on land area. However, they must cover the land, and therefore are much more intrusive. I have guessed that 0.06 % of land area may be covered, which adds 380 W/person to the ledger.

Nuclear capability is decreasing in the United States and the industry remains under pressure in Europe. I nevertheless assume a modest nuclear renaissance and allow a tripling of the power output. This is likely about all that can be achieved because the global uranium mining industry is constrained by politics and supply.

This leaves the great uncertainty in the projections: fusion power. Unfortunately, it is not clear that viable fusion power is possible. Certainly it is at least 30 years away. Nevertheless, carbon-dependence must be eliminated as soon as possible, and uncertainty over fusion power should not be used as an excuse. Even if fusion reactors do not prove feasible, Table 10.1 indicates that about one half of the current global average energy use can be generated in a renewable and Earth-friendly manner. If transportation is electrified and waste is reduced it should be possible to get very close to the goal.

Clearly fossil fuels carry more costs to society than their simple removal from the ground. These costs should be included in their price, either by a tax or some other mechanism. Revenues generated should then be invested in developing more efficient PV cells, electrical power grids, coal sequestration technology, fusion energy, breeder reactors, and efficient batteries. It is certain that the government will have to take the lead in this research and development because market forces are not sufficient to lead the way.

Our discussion makes it clear that political and societal resolve needs to be strong to make these things happen. Certainly, the population will need to stabilize at some sustainable value – likely not much higher than it is now, and possibly much lower. As we have seen this is predicted to happen naturally, but it is preferable that governments assist the process with educational and health care programs.[12] Perhaps most importantly, it will require societal pressure to migrate the current global paradigm of growth-capitalism to the required future of sustainable-capitalism.

Managing a finite planet is a global shared responsibility. It is not the time for misguided or misinformed thinking. Think back to the first three chapters of this book where we studied what science is, how it is done, and how it can be abused. If you are voting or acting on an issue based on the ideas of orgonomy or qi, or because you do not understand the importance of energy flow in the environment, then you are not helping society grow in useful directions. The overriding concern is that pseudoscience and the muddle-headed thinking that accompanies it disempowers people. We must strive to resist this trend. Hopefully this book will help you make your personal decisions wisely. Above all, remember to query your information and assumptions!

Ask questions.

---

[12]The educational level of women is known to be strongly negatively correlated with family size. Of course, a robust affordable health care system also reduces pressure on reproduction rates.

REVIEW

Important terminology:

electrolysis   [pg. 242]

glucose oxidation   [pg. 242]

ozone, CFC   [pg. 254]

propellant   [pg. 246]

rocket equation   [pg. 246]

spacetime   [pg. 248]

Important concepts:

Difficulties with short range space travel: radiation, muscle and bone loss, expense.

Difficulties with long range space travel: energy requirements, time requirements.

Difficulties with colonization: harsh environment, energy requirements, infrastructure requirements.

The concepts of space and time are relative to motion.

Market forces are not sufficient to implement a sustainable future.

Sustainability will require societal action, a price on carbon, and investment in new infrastructure.

Transition from a growth economy to a sustainable economy must be achieved.

FURTHER READING

J. Diamond, *Guns, Germs, and Steel: The Fates of Human Societies*, W.W. Norton & Company, 1999.

R. Wright, *A Short History of Progress*, Da Capo Press, 2005.

EXERCISES

1. Rocket Power.

   In Sect. 10.1.3 the claim was made that the minimum power required to a spacecraft was $1000 f^3$ GW/person. Verify this.

2. Rocket Equation.

   (a) Obtain the rocket equation by explictly conserving momentum.

   (b) Obtain Eq. 10.5 by conserving energy.

   The following figure should help set up the required equations.

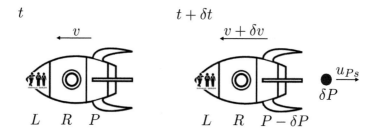

3. Energy Density.

   Compute the amount of energy produced (in joules) if a $\frac{1}{2}$ kg of matter annihilates with a $\frac{1}{2}$ kg of antimatter. Could this energy source power a spaceship? Refer to Sect. 4.1 for help.

4. Fast Motion.

   Einstein's theory is important for fast motion.

   (a) Argue that Einsteinian equations should reduce to Newtonian (i.e., slow) equations when c is large.

   (b) Examine Eqs. 10.10 and 10.11 and consider the limit as c becomes large. Are the expectations of (a) obtained?

5. Space Travel.

   (a) Given the constraints discussed in this chapter, do you think that interstellar space travel of any sort is possible?

   (b) Do you think it will be possible for humans to live on Mars at any time in the future? What obstacles must be overcome?

6. Future Humanity

   (a) What does it mean to be human? Would it be possible for humanity to exist in a non-biological form?

   (b) What implications does this have for space travel? For society?

7. Future Growth.

   (a) Will economic growth continue in the future?

   (b) Will population growth continue in the future?

   (c) Will scientific knowledge continue to grow in the future?

8. Free Market Forces.

   Discuss whether the free market is capable of transitioning to a sustainable economy.

9. Carbon Tax

   In Sect. 10.2.3 it was stated that carbon should be taxed. Do you agree with this? What other mechanisms to account for the cost of carbon can be employed?

10. Recycling

    In Sect. 10.2.3 it was claimed that the free market should be able to transition from mining resources to recycling resources. Do you agree with this? In what ways might this transition require regulatory help?

11. Future Energy.

    Come up with a scenario where global energy requirements can be met by non-carbon sources. Assume that fusion does not pan out. What compromises will have to be made?

12. GMO Foods.

    Read about genetically modified foods. Where do your sympathies lie in this controversy? Do you anticipate problems with unforeseen consequences? Is this progress as usual, or has science overstepped some boundary?

13. Autonomous Cars.

    It is likely that driverless cars will become a reality in the coming decades. What changes will this new technology bring about? Will it be necessary to own a car?

Problem Solutions

The solutions to a selection of representative quantitative problems are presented here to assist students who may need guidance.

Alpha, Beta, and True Hypotheses

A researcher with a lot of time examines 1000 hypotheses of which 100 are true. Assume alpha = 0.05 and beta = 0.79 (this corresponds to a power of 0.21, which is typical of neuroscience studies). What fraction of "true" hypotheses are actually true?

Alpha is the odds of making a type I error, which is the fraction of false hypothesis that are mistaken as true. Since there are 900 false hypotheses, the number of false positive results is given by n where

$$\text{alpha} = 0.05 = \frac{n}{900}, \; n = 0.05 \cdot 900 = 45.$$

Beta is the odds of obtaining a false negative, which is the ratio of true hypotheses that are mistaken as false. Since there are 100 true hypotheses, there are m false negatives, where

$$\text{beta} = 0.79 = \frac{m}{100}, \; m = 0.79 \cdot 100 = 79.$$

This leave $100 - 79 = 21$ true hypotheses that have been identified as true. But the researcher has identified $21 + 45 = 66$ hypotheses as being true. Of these only $21/66 = 31.8\%$ are actually true.

P-Values and Hypothesis Testing

You wish to test the hypothesis that a coin is fair (i.e., the odds of coming up heads is ½). You flip the coin six times and obtain 5 heads.

(a) Compute the probability of obtaining at least 5 heads.

© Springer International Publishing Switzerland 2016
E.S. Swanson, *Science and Society*,
DOI 10.1007/978-3-319-21987-5

(b) Take the null hypothesis to be that the coin is fair. If your test criterion is alpha <0.05, do you accept or reject the null hypothesis?

(c) What p-value would you assign to the statement that the coin is fair?

> (a) *The probability of obtaining any particular combination of six heads and tails is*
>
> $$\frac{1}{2} \cdot \frac{1}{2} \cdots \cdot \frac{1}{2} = \frac{1}{2^6}.$$
>
> *There are six ways to get five heads in six throws:*
>
> *thhhhh, hthhhh, hhthhh, hhhthh, hhhhth, hhhhht.*
>
> *Thus the odds of getting five or six heads in six tosses is*
>
> $$\frac{6+1}{2^6} = 0.109.$$

> (b) *The null hypothesis is accepted since 0.109 > 0.05. I.e., the odds of obtaining five or six heads is larger than the selected value for alpha, thus there is not sufficient evidence to conclude the coin is not fair.*
>
> (c) *The p-value is the probability of obtaining the results seen or greater, given that the null hypothesis is true. In this case, it is the odds of obtaining five or six heads assuming the coin is fair. This is the number computed in part (a), hence p = 0.109.*

Energy Cost

Estimate the cost of air conditioning the Consol Energy Center for 3 h. In Pittsburgh electricity sells for 10 ¢/kWh.

This is a problem in converting units. Assume that the rink holds 20,000 people, who produce heat at a rate of 2000 Cal/day. This is (Chap. 4)

$$2000 \frac{\text{Cal}}{\text{day}} \times 4184 \frac{\text{J}}{\text{Cal}} \times \frac{1}{24 \times 60 \times 60} \frac{\text{day}}{\text{s}} = 96.7 \text{ W}.$$

We assume that all this energy must be removed by the air conditioning unit, and that there is no waste in the system. The total energy produced in three hours is

$$20000 \cdot 3 \text{ hr} \cdot 3600 \frac{\text{s}}{\text{hr}} \cdot 96.7 \frac{\text{J}}{\text{s}} = 20.9 \text{ GJ}.$$

Energy is supplied at 10 ¢/kWh, and a kilowatt-hour is

$$1 \text{ kWh} = 1000 \frac{\text{J} \cdot \text{hr}}{\text{s}} \cdot 3600 \frac{\text{s}}{\text{hr}} = 3.6 \text{ MJ}.$$

Thus the cost to air condition the rink for three hours is approximately

$$20.9 \text{ GJ} \cdot \frac{\$0.1}{3.6 \text{ MJ}} = \$580.$$

Atomic Transition Energy, Wavelength and Frequency

Compute the energy, wavelength, and frequency of a photon that is emitted when hydrogen makes a transition from the fifth to the second orbital.

This problem can be solved with Rydberg's formula, Eq. 5.6, which gives the frequency of light emitted when a hydrogen atom makes a transition.

$$f = R\left(\frac{1}{n^2} - \frac{1}{m^2}\right) = 3.28 \cdot 10^{15}\left(\frac{1}{2^2} - \frac{1}{5^2}\right) \text{ Hz} = 6.89 \cdot 10^{14} \text{ Hz}.$$

Now use $\lambda = c/f$ to obtain the photon wavelength:

$$\lambda = \frac{2.99 \cdot 10^8}{6.89 \cdot 10^{14}} = 4.34 \cdot 10^{-7} \text{ m}.$$

The energy of the photon is given by the Planck-Einstein formula

$$E = hf = 6.626 \cdot 10^{-34} \cdot 6.89 \cdot 10^{14} = 4.565 \cdot 10^{-19} \text{ J}.$$

Divide by $1.602 \cdot 10^{-19}$ to obtain this energy in electron volts:

$$E = \frac{4.565 \cdot 10^{-19} \text{ J}}{1.602 \cdot 10^{-19} \text{ J/eV}} = 2.850 \text{ eV}.$$

Dimensional Analysis and Planck's Formula

Verify the units in Planck's formula for blackbody radiation, Eq. 6.1:

$$U = \frac{2hf^3}{c^2} \frac{1}{e^{hf/k_B T} - 1}.$$

First, notice that the denominator in the second factor must have no units because the number one has no units. The only way this can happen is if the argument of the exponential also has no units (e.g., one never sees an expression like $\exp(L)$, where L is a length or anything else with units). Because the units of hf are energy (J), the units of the Boltzmann constant must be J/K (i.e., joules per unit temperature). The units of U must therefore be given by the first factor, which is

$$\frac{[hf][f^2]}{[c^2]} = \frac{J(1/s^2)}{(m/s)^2} = \frac{J}{m^2}.$$

This is fine, but usually the units for Planck's formula are expressed in terms of power. Since the units of power are J/s, one needs to rewrite the latter expression as

$$\frac{Ws}{m^2} = \frac{W}{Hz \cdot m^2}.$$

Lastly, Planck's derivation of his formula was in terms of "unit solid angle", which is an angular measure of a surface area, and carries no units.

Skin Temperature and the Stefan-Boltzmann Law

Use the Stefan-Boltzmann law and a human power of 97 W to estimate skin temperature at an ambient temperature of 23 degrees C.

Skin will receive energy from two sources: the body and the surrounding air. Thus, in terms of power

$$P_{body} + P_{air} = P_{skin},$$

We assume that the energy flow for air and skin can be described by the Stefan-Boltzmann law, Eq. 6.2. Thus $P_{air} = A\sigma T_{air}^4$ and $P_{skin} = A\sigma T_{skin}^4$, where A is the surface area of a person. Substitute to obtain

$$T_{skin}^4 = T_{air}^4 + \frac{P_{body}}{A\sigma}.$$

Take $A = 2\,m^2$, $T_{air} = 296\,K$, and $P_{body} = 97\,W$ to obtain

$$T_{skin} = 303.9\,K = 30.8\,C.$$

This figure is pretty close to the measured average skin temperature of 31.6 C.

Atomic Recoil

An atom of mass $1200\,MeV/c^2$ is at rest when it absorbs a photon of frequency 1.4 THz. It then emits a photon of frequency 0.8 THz. Assuming that it starts and ends in the ground state (n = 1), how fast must the atom be moving after emitting the photon?

The solution to this problem is based on conservation of energy; specifically Eq. 6.9. Because the atom starts and ends in the same electronic configuration, $E_{electrons}$ can be ignored. Solving gives

$$E_{\substack{atomic \\ motion}}\Big|_{after} = 0 + hf\big|_{before} - hf\big|_{after},$$

where the "0" appears because the atom is initially at rest. Thus

$$E_{\substack{\text{atomic} \\ \text{motion}}}\Big|_{\text{after}} = h(1.4 \cdot 10^{12} - 0.8 \cdot 10^{12}) \text{ Hz} = 3.98 \cdot 10^{-22} \text{ J}.$$

The speed of the atom can now be obtained using

$$E_{\substack{\text{atomic} \\ \text{motion}}}\Big|_{\text{after}} = KE = \frac{1}{2}mv^2.$$

We will need the mass of the atom in SI units. The conversion factor is given in Eq. 4.5, and we obtain

$$1200 \text{ MeV/c}^2 = 1.2 \cdot 1.783 \cdot 10^{-27} = 2.14 \cdot 10^{-27} \text{ kg}.$$

Thus

$$v = \sqrt{\frac{2 \cdot KE}{m}} = \sqrt{\frac{2 \cdot 3.98 \cdot 10^{-22}}{2.14 \cdot 10^{-27}}} = 610 \text{ m/s}.$$

Slow Fission Products

In Sect. 8.5.3 it was argued that fission products move at 6 % of the speed of light.

This figure is obtained as follows. Simplify the problem by assuming that uranium-235 splits into equal fragments. The mass of one fragment is thus

$$\frac{1}{2} \cdot 235 \cdot 938 \text{ MeV/c}^2 = 110 \text{ GeV/c}^2$$

where we quote the mass of a proton or neutron in electron volts. The kinetic energy of the fission products is about 200 MeV, and this is given by the equation

$$KE = \frac{1}{2}mv^2.$$

Thus

$$v = \sqrt{\frac{2 \cdot KE}{m}} = \sqrt{\frac{2 \cdot 200}{110000}} \, c = 0.06c.$$

Thus the speed of a fission fragment is about 6 % of the speed of light. A more accurate calculation would use Eq. 8.9, but this makes little difference.

LNT and Cancer

It is estimated that 2 million people received average doses of 14 µSv due to the TMI incident. Evaluate the additional number of cancer cases this caused using the LNT model.

We know that 1 Sv is equivalent to a 5.5 % chance of getting cancer. Thus 14 µSv corresponds to $0.055 \cdot 14 \cdot 10^{-6} = 0.77 \cdot 10^{-6}$ chance of cancer. This is for each of 2 million people, thus $2 \cdot 10^6 \cdot 0.77 \cdot 10^{-6} = 1.54$ extra incidents of cancer will occur.

Exponential Growth of Wind Power

Figure 9.15 indicates that global wind power capacity is increasing exponentially at 21 %/yr. When does capacity reach 16 TW?

Capacity is 318 GW in 2013. Thus the formula describing the growth of wind power is

$$P = P_0 \cdot \exp(r(t - t_0)),$$

with $P_0 = 318\,GW$, $t_0 = 2013\,yr$, and $r = 0.21$. To find when this hits 16 TW, we set $P = 16{,}000\,GW$ and solve for t:

$$\frac{P}{P_0} = \exp(r(t - t_0)).$$

Take a logarithm:

$$\ln\left(\frac{P}{P_0}\right) = r(t - t_0),$$

$$t - t_0 = \frac{1}{r}\ln\left(\frac{P}{P_0}\right),$$

$$t = t_0 + \frac{1}{r}\ln\left(\frac{P}{P_0}\right).$$

Plugging in the values above gives $t = 2013 + 18.66 = 2031.66\,yr$.

Index

© Springer International Publishing Switzerland 2016
E.S. Swanson, *Science and Society*,
DOI 10.1007/978-3-319-21987-5